VIDEO ORGANIZER

MATH MADE VISIBLE LLC

BEGINNING ALGEBRA

SEVENTH EDITION

Elayn Martin-Gay

University of New Orleans

PEARSON

Boston   Columbus   Indianapolis   New York   San Francisco

Amsterdam   Cape Town   Dubai   London   Madrid   Milan   Munich   Paris   Montreal   Toronto

Delhi   Mexico City   São Paulo   Sydney   Hong Kong   Seoul   Singapore   Taipei   Tokyo

The author and publisher of this book have used their best efforts in preparing this book. These efforts include the development, research, and testing of the theories and programs to determine their effectiveness. The author and publisher make no warranty of any kind, expressed or implied, with regard to these programs or the documentation contained in this book. The author and publisher shall not be liable in any event for incidental or consequential damages in connection with, or arising out of, the furnishing, performance, or use of these programs.

Reproduced by Pearson from electronic files supplied by the author.

ISBN-13: 978-0-13-421671-3
ISBN-10: 0-13-421671-7

www.pearsonhighered.com

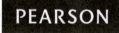

CONTENTS

Section 1.1 Study Skill Tips for Success in Mathematics

Complete the outline as you view Lecture Video 1.1. Pause ⏸ the video as needed as you fill in all blanks. Circle your answer to each numbered exercise. Then press Play ▶ to continue listening to the video.

Objective A **Get ready for this course.**

Have a positive attitude!

Check to see that you understand the way that this course is taught.

On the first day of class, bring all the materials.

Objective B **Understand some general tips for success.**

⏸ Stay _____!

⏸ Attend all _____ and be on _____.

Get help as soon as you need it.

⏸ Learn from your _____.

Turn in assignments on time.

Objective C **Know how to use this text.**

⏸ Open your _____ and become familiar with it.

⏸ The pencil symbol means a _____ _____ _____ exercise.

⏸ The triangle symbol means the exercise has something to do with _____.

The ▶ means that the corresponding exercise can be viewed on the DVD lecture series.

Section 1.1 Study Skill Tips for Success in Mathematics

Objective D   **Know how to use text resources.**

The two resources focused on are _____ Resources and _____ Resources.

DVD Lecture Videos correspond to every section in the book.

The _____ _____ Prep Videos contain the solutions to all the Chapter Test exercises.

Student Success Tips are 3 minute reminders.

The _____ _____ Videos are videos that show the solutions to every exercise in the final practice exam.

_____ Organizer follows the DVD Lecture Series.

The_____ Organizer contains tips for note-taking, practice exercises before homework, and references to the text and DVD lecture series.

Objective E   **Get help as soon as you need it.**

Get help with mathematics as soon as you need it.

Objective F   **Learn how to prepare for and take an exam.**

When you think you are ready, take a _____ _____.

Remember the Chapter Test Prep Videos contain the worked out solutions to all end-of-chapter Chapter Test Exercises.

Objective G   **Develop good time management.**

Try filling out the Time Grid at the end of Section 1.1.

Section 1.2 Symbols and Sets of Numbers

Complete the outline as you view Video Lecture 1.2. Pause ⏸ the video as needed to fill in the blanks. Then press Play ▶ to continue. Also, circle your answer to each numbered exercise.

(**Objective 1**) **Use a number line to order numbers**

A set is a collection of objects called members or elements.

⏸ **Common Sets of Numbers**

Natural Numbers: The set of natural numbers is _____.

Whole Numbers: The set of whole numbers is _____.

⏸ **Review of Symbols**

The symbol "=" means _____.

The symbol "≠" means _____.

The symbol "<" means _____.

The symbol ">" means _____.

▶ **Work Video Exercises 1 and 2 with me.**

Insert <, >, or = to form a true statement.

▶ 1. 7   3   ▶ 2. 0   7

⏸ The symbol "≤" means _____.

The symbol "≥" means _____.

▶ **Work Video Exercise 3 with me.**

3. Is $11 \leq 11$ a true or false statement?

3

Section 1.2 Symbols and Sets of Numbers

(Objective 2) **Translate sentences into mathematical statements**

⊙ **Work Video Exercise 4 with me.**

4. Write as a mathematical statement: five is greater than or equal to four.

Now press pause ⏸ **and work Video Exercise 5. Then press Play** ⊙ **and check your work.**

5. Write as a mathematical statement: fifteen is not equal to negative two.

⊙ Play and check.

(Objective 3) **Identify natural numbers, whole numbers, integers, rational numbers, irrational numbers, and real numbers**

⏸ **Common Sets of Numbers**

   Integers: The set of integers is _____.

   Rational Numbers: The set of rational numbers is _____.

   Irrational Numbers: The set of irrational numbers is _____.

   Real Numbers: The set of real numbers is _____.

⊙ **Work Video Exercise 6 with me.**

6. Tell which set(s) 0 belongs to.

   _____ Natural Numbers

   _____ Whole Numbers

   _____ Integers

   _____ Rational Numbers

   _____ Irrational Numbers

   _____ Real Numbers

Section 1.2 Symbols and Sets of Numbers

Now press pause ⏸ **and work Video Exercise 7. Then press Play** ▶ **and check your work.**

7. Tell which set(s) $\dfrac{2}{3}$ belongs to.

_____ Natural Numbers

_____ Whole Numbers

_____ Integers

_____ Rational Numbers

_____ Irrational Numbers

_____ Real Numbers

▶ Play and check.

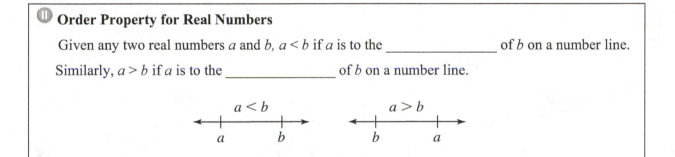

⏸ **Order Property for Real Numbers**

Given any two real numbers a and b, $a < b$ if a is to the _____ of b on a number line.

Similarly, $a > b$ if a is to the _____ of b on a number line.

$a < b$ $a > b$

 a   b   b   a

Objective 4 **Find the absolute value of a real number**

⏸ _____ _____ is the distance between a real number a and 0 on a number line.

▶ **Work Video Exercise 8 with me.**

Insert <, >, or = to form a true statement.

8. $|-5|$ -4

Section 1.2 Symbols and Sets of Numbers

Now press pause ⏸ and work Video Exercise 9. Then press Play ▶ and check your work.

Insert <, >, or = form a true statement.

9. $|0|$ $|-8|$

▶ Play and check.

Section 1.3 Fractions and Mixed Numbers

Complete the outline as you view Video Lecture 1.3. Pause ⏸ the video as needed to fill in the blanks. Then press Play ▶ to continue. Also, circle your answer to each numbered exercise.

Objective 1 **Write fractions in simplest form**

⏸ **Prime Number**
A prime number is a _____ _____ other than 1, whose only factors are _____ and _____. The first few prime numbers are 2, 3, 5, 7, 11, 13, 17, 19, 23, 29, and so on.

⏸ A _____ _____ other than 1 that is not a prime number is a _____ _____.

Fundamental Principle of Fractions
If $\dfrac{a}{b}$ is a fraction and c is a non-zero real number, then $\dfrac{a \cdot c}{b \cdot c} = \dfrac{a}{b}$.

$\dfrac{c}{c} = 1$, where c is a real number, $c \neq 0$

▶ **Work Video Exercise 1 with me.**

Simplify.

1. $\dfrac{10}{15} = \dfrac{\underline{} \cdot \underline{}}{\underline{} \cdot \underline{}}$

   $= \underline{}$

Objective 2 **Multiply and divide fractions**

Multiplying Fractions
$\dfrac{a}{b} \cdot \dfrac{c}{d} = \dfrac{a \cdot c}{b \cdot d}$ if $b \neq 0$ and $d \neq 0$

⏸ **To multiply fractions:** Multiply the _____ and multiply the _____.

Section 1.3 Fractions and Mixed Numbers

⏵ **Work Video Exercise 2 with me.**

Simplify.

2. $\dfrac{2}{3} \cdot \dfrac{3}{4} = \dfrac{\underline{\quad} \cdot \underline{\quad}}{\underline{\quad} \cdot \underline{\quad}}$

$= \dfrac{\underline{\qquad}}{}$

$= \dfrac{\underline{\qquad}}{}$

Dividing Fractions

$\dfrac{a}{b} \div \dfrac{c}{d} = \dfrac{a}{b} \cdot \dfrac{d}{c}$ if $b \neq 0$, $d \neq 0$, and $c \neq 0$

⏵ **Work Video Exercise 3 with me.**

Simplify.

3. $\dfrac{3}{4} \div \dfrac{1}{20} = \dfrac{\underline{\quad}}{\underline{\quad}} \cdot \dfrac{\underline{\quad}}{\underline{\quad}}$

$= \dfrac{\underline{\quad} \cdot \underline{\quad}}{\underline{\quad} \cdot \underline{\quad}}$

$= \underline{\quad}$ or $\underline{\quad}$

Adding and Subtracting Fractions with the Same Denominator

$\dfrac{a}{b} + \dfrac{c}{b} = \dfrac{a+c}{b}$, if $b \neq 0$

$\dfrac{a}{b} - \dfrac{c}{b} = \dfrac{a-c}{b}$, if $b \neq 0$

Section 1.3 Fractions and Mixed Numbers

Objective 3 **Add and subtract fractions**

 **Work Video Exercise 4 with me.**

Subtract and simplify.

4. $\dfrac{17}{21} - \dfrac{10}{21} = \dfrac{\underline{} - \underline{}}{\underline{}}$

$$= \dfrac{}{} = \dfrac{}{}$$

$$= \underline{}$$

⏸ A fraction is _____ when the numerator and denominator have no common factors other than _____ or _____ .

⏸ _____ _____ are equal and represent the same quantity.

 **Work Video Exercise 5 with me.**

Write an equivalent fraction.

5. $\dfrac{4}{5} = \dfrac{\cdot}{\cdot} = \underline{}$

 **Work Video Exercise 6 with me.**

Subtract.

6. $\dfrac{10}{3} - \dfrac{5}{21} = \dfrac{10\cdot\underline{}}{3\cdot\underline{}} - \dfrac{5}{21}$ LCD = _____

$$= \dfrac{\underline{}}{\underline{}} - \dfrac{5}{21}$$

$$= \underline{}$$

Section 1.3 Fractions and Mixed Numbers

(Objective 4) **Perform operations on mixed numbers**

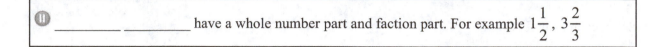

$\quad$ _____ _____ have a whole number part and faction part. For example $1\frac{1}{2}$, $3\frac{2}{3}$

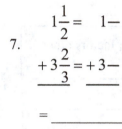

Work Video Exercise 7 with me.

Add.

7. $\quad 1\frac{1}{2} = \quad 1—$

$\quad +3\frac{2}{3} = +3—$ _____

$= \underline{\hspace{2cm}} + \underline{\hspace{2.5cm}} = \underline{\hspace{2cm}} + \underline{\hspace{2cm}}$

$= \underline{\hspace{2cm}} + \underline{\hspace{2.5cm}} + \underline{\hspace{2cm}} = \underline{\hspace{2cm}}$

_____ _____ : Numerator greater than or equal to denominator

Section 1.4 Exponents, Order of Operations, Variable Expressions, and Equations

Complete the outline as you view Video Lecture 1.4. Pause ⏸ the video as needed to fill in the blanks. Then press Play ▶ to continue. Also, circle your answer to each numbered exercise.

Objective 1 **Define and use exponents and order of operations**

⏸ **Exponential Notation**
A short-hand notation for _____ _____ of the same _____.

⏸ Evaluate means "_____".

▶ **Work Video Exercise 1 with me.**

Evaluate.

1. Three to the third power

Now press Pause ⏸ and work Video Exercise 2. Then press Play ▶ and check your work.

Evaluate.

2. $\left(\dfrac{2}{3}\right)^4$

▶ Play and check.

Order of Operations
Simplify expressions using the order below. If grouping symbols such as parentheses are present, simplify expressions within those first, starting with the innermost set. If fraction bars are present, simplify the numerator and the denominator separately.

⏸ **Order of Operations**
Step 1: _____ exponential expressions.
Step 2: Perform _____ or _____ in order from left to right.
Step 3: Perform _____ or _____ in order from left to right.

Section 1.4 Exponents, Order of Operations, Variable Expressions, and Equations

▶ **Work Video Exercise 3 with me.**

Evaluate.

3. $2\left[5+2(8-3)\right]$

⏸ With no grouping symbols, _____ or _____ before _____ or _____ .

Now press Pause ⏸ **and work Video Exercise 4. Then press Play** ▶ **and check your work.**

Evaluate.

4. $\dfrac{|6-2|+3}{8+2\cdot 5}$

▶ Play and check.

(**Objective 2**) **Evaluate algebraic expressions, given replacement values for variables**

⏸ A symbol used to represent an unknown number is called a _____ .

Algebraic expression: contains numbers, variables, operation symbols and grouping symbols.

▶ **Work Video Exercise 5 with me.**

5. If $x=1$, $y=3$ and $z=5$, evaluate $|2x+3y|$.

⏸ Absolute value bars can be _____ _____ .

Section 1.4 Exponents, Order of Operations, Variable Expressions, and Equations

Now press Pause ⏸ **and work Video Exercise 6. Then press Play** ▶ **and check your work.**

6. If $x = 12$, $y = 8$, and $z = 4$, evaluate $\dfrac{x^2 + z}{y^2 + 2z}$.

▶ Play and check.

⏸ Fraction bars can be _____ _____.

Objective 3 **Determine whether a number is a solution of a given equation**

⏸ An <u>equation</u> is of the form: _____ = _____.

▶ **Work Video Exercise 7 with me.**

7. Is 6 a solution of $3x - 10 = 8$?

A <u>solution</u> of an equation is a value for the variable that makes the equation true.

Now press Pause ⏸ **and solve Video Exercise 8. Then press Play** ▶ **and check your work.**

Decide whether the given number is a solution of the equation.

8. Is 0 a solution of $x = 5x + 15$?

▶ Play and check.

Section 1.4 Exponents, Order of Operations, Variable Expressions, and Equations

(Objective 4) **Translate phrases into expressions and sentences into statements**

⏵ **Work Video Exercise 9 with me.**

Write the phrase as an algebraic expression.

9.   Three times a number, increased by 22

⎯⎯⎯⎯⎯⎯⎯⎯⎯⎯⎯⎯⎯⎯⎯⎯⎯⎯⎯⎯⎯⎯⎯⎯⎯⎯⎯⎯⎯⎯

⏸ An equation contains _____, an _____ does not.

⎯⎯⎯⎯⎯⎯⎯⎯⎯⎯⎯⎯⎯⎯⎯⎯⎯⎯⎯⎯⎯⎯⎯⎯⎯⎯⎯⎯⎯⎯

Now press Pause ⏸ and work Video Exercises 10 and 11. Then press Play ⏵ and check your work.

Write each sentence as an equation or inequality.

10.   One increased by two equals the quotient of nine and three.

⏵ Play and check.

11.   Three is not equal to four divided by two.

⏵ Play and check.

Section 1.5 Adding Real Numbers

Complete the outline as you view Video Lecture 1.5. Pause ⏸ the video as needed to fill in the blanks. Then press Play ▶ to continue. Also, circle your answer to each numbered exercise.

Objective 1 Add real numbers

⏸ _____ _____ = {All numbers that correspond to points on the number line.}

Adding Two Numbers with the Same Sign: Add their absolute values. Use their common sign as the sign of the sum.

▶ **Work Video Exercise 1 with me.**

Add.

1. $-6+(-8)$

Now press Pause ⏸ and work Video Exercises 2, 3, and 4. Then press Play ▶ and check your work.

2. $-2+(-3)$

▶ Play and check.

3. $-9+(-3)$

▶ Play and check.

4. $-21+(-16)+(-22)$

▶ Play and check.

Adding Two Numbers with Different Signs: Subtract the smaller absolute value from the larger absolute value. Use the sign of the number whose absolute value is larger as the sign of the sum.

▶ **Work Video Exercise 5 with me.**

Add.

5. $5+(-7)$

Section 1.5 Adding Real Numbers

Now press Pause ⏸ and work Video Exercises 6, 7, and 8. Then press Play ▶ and check your work.

Add.

6. $6.3 + (8.4)$

▶ Play and check.

7. $|-6| + (61)$

▶ Play and check.

8. $6 + (-4) + 9$

▶ Play and check.

⏸ **Order of Operations**
    Add or subtract from _____ to_____.

Objective 2 **Solve applications that involve addition of real numbers**

▶ **Work Video Exercise 9 with me.**

9.   The low temperature in Anoka, Minnesota, was $-15°$ last night. During the day it rose only $9°$. Find the high temperature for the day.

Objective 3 **Find the opposite of a number**

Opposites or Additive Inverses
Two numbers that are the same distance from 0 but lie on opposite sides of 0 are called opposites or additive inverses of each other.

Section 1.5 Adding Real Numbers

Work Video Exercise 10 with me.

Find the opposite or additive inverse.

10.   6
      opposite:

      Now add the number 6 and its opposite: $6 + (-6)$

Now press Pause ⏸ **and work Video Exercise 11. Then press Play** ▶ **and check your answer.**

Find the opposite or additive inverse of -2.

11. -2
      opposite:

      Find the sum: $-2 + 2 =$

▶ Play and check.

The sum of a number a and its opposite, $-a$, is 0.
   $$a + (-a) = 0$$

⏸ "The opposite of" translates to "_____".

If a is a number, then $-(-a) = a$.

Work Video Exercises 12 and 13 with me.

Simplify each expression.

12. $-(-7)$ 13. $-|-2|$

Section 1.5 Adding Real Numbers

Complete the outline as you view Video Lecture 1.6. Pause ⏸ the video as needed to fill in the blanks. Then press Play ▶ to continue. Also, circle your answer to each numbered exercise.

⬭ **Objective 1** **Subtract real numbers**

⏸ **Subtracting Two Real Numbers**
   If a and b are real numbers, then $a - b =$ _____ + _____.

▶ **Work Video Exercise 1 with me.**

Subtract.

1. $16 - (-3)$

Now press Pause ⏸ **and work Video Exercises 2 and 3. Then press Play** ▶ **and check your work.**

2. $-6 - 5$ 3. $\dfrac{-3}{11} - \left(\dfrac{-5}{11}\right)$

▶ Play and check.   ▶ Play and check.

▶ **Work Video Exercise 4 with me.**

Translate the phrase into an expression, then simplify.

4.   Subtract 9 from -4.

⬭ **Objective 2** **Add and subtract real numbers**

Now press Pause ⏸ **and work Video Exercise 5. Then press Play** ▶ **and check your work.**

Simplify.

5. $-10 - (-8) + (-4) - 20$

▶ Play and check.

⏸ **Remember with Order of Operations:** Add or subtract from _____ to _____.

Section 1.6 Subtracting Real Numbers

Work Video Exercise 6 with me.

Simplify, following the order of operations.

6. $|-3| + 2^2 + \left[-4 - (-6)\right]$

Objective 3 Evaluate algebraic expressions using real numbers

Work Video Exercise 7 with me.

7. Evaluate when $x = -5$ and $y = 4$.
$$\frac{9 - x}{y + 6}$$

Objective 4 Solve applications that involve of real numbers

An equation is of the form: _____ = _____

Now press Pause ⏸ **and work Video Exercise 8. Then press Play** ▶ **and check your answer.**

8. A commercial jet liner hits an air pocket and drops 250 feet. After climbing 120 feet, it drops another 178 feet. What is its overall vertical change?

Play and check.

Objective 5 Find complementary and supplementary angles

Complementary and Supplementary Angles
Two angles are _____ if their sum is $90°$.
Two angles are _____ if their sum is $180°$

Section 1.6 Subtracting Real Numbers

▶ **Work Video Exercise 9 with me.**

Find the measure of angle *y*.

9.

Now press Pause ⏸ **and work Video Exercise 10. Then press Play** ▶ **and check your work.**

Find the measure of angle *x*.

10.

▶ Play and check.

Section 1.6 Subtracting Real Numbers

Section 1.7 Multiplying and Dividing Real Numbers

Complete the outline as you view Video Lecture 1.7. Pause ⏸ the video as needed to fill in the blanks. Then press Play ▶ to continue. Also, circle your answer to each numbered exercise.

(**Objective 1**) **Multiply real numbers**

⏸ An _____ is any number that is to be added.

▶ **Work Video Exercise 1 with me.**

Multiply.

1. $3 \cdot (-2)$

⏸ The product of two numbers with different signs is a _____ _____.

⏸ The product of two numbers with the same signs is a _____ _____.

Now press Pause ⏸ and work Video Exercises 2 and 3. Then press Play ▶ and check your work.

Multiply.

2. $-6(4)$ 3. $2(-1)$

▶ Play and check.   ▶ Play and check.

▶ **Work Video Exercise 4 with me.**

4. Multiply $-5(-10)$.

Now press Pause ⏸ and work Video Exercise 5. Then press Play ▶ and check your work.

5. Multiply $\left(\dfrac{2}{3}\right)\left(\dfrac{-4}{9}\right)$.

▶ Play and check.

Section 1.7 Multiplying and Dividing Real Numbers

▶ **Work Video Exercises 6 and 7 with me.**

Evaluate.

6. $(-2)^4$

7. -2^4

Without parentheses, such as -2^4, the exponent 4 applies to the base of 2 only.

⏸ **Multiplying by Zero**
If b is a real number, then $b \cdot 0 =$ _____ . Also $0 \cdot b =$ _____ .

▶ **Work Video Exercise 8 with me.**

Multiply.

8. $-7 \cdot 0$

Reciprocals or Multiplicative Inverses
Two numbers whose product is 1 are called reciprocals or multiplicative inverses of each other.

⬭**Objective 2** **Find the reciprocal of a real number**

▶ **Work Video Exercise 9 with me.**

Find the reciprocal.

9. $\dfrac{2}{3}$

Now press Pause ⏸ and work Video Exercise 10. Then press Play ▶ and check your work.

10. Find the reciprocal of -14.

▶ Play and check.

⏸ The reciprocal of $\dfrac{a}{b}$ is _____ .

Section 1.7 Multiplying and Dividing Real Numbers

Quotient of Two Real Numbers

If a and b are real numbers and $b \neq 0$, then $a \div b = \dfrac{a}{b} = a \cdot \dfrac{1}{b}$.

Multiplying and Dividing Real Numbers
1. The product _or_ quotient of two numbers with the same sign is a _____ number.
2. The product _or_ quotient of two numbers with different signs is a _____ number.

Objective 3 Divide real numbers

▶ **Work Video Exercise 11 with me.**

Divide.

11. $\dfrac{18}{-2}$

Now press Pause ⏸ and work Video Exercise 12. Then press Play ▶ and check your work.

Divide.

12. $\dfrac{-12}{-4}$

▶ Play and check.

▶ **Work Video Exercises 13, 14, and 15 with me. Pause ⏸ as needed to fill in the blanks in the boxes on the right.**

Divide.

13. $\dfrac{-5}{9} \div \left(\dfrac{-3}{4} \right)$

⏸ Divide Fractions: $\dfrac{a}{b} \div \dfrac{c}{d} = \dfrac{a}{b} \cdot$ _____

14. $\dfrac{0}{-4}$

⏸ If $a \neq 0$, $\dfrac{0}{a}$ is _____.

15. $\dfrac{5}{0}$

⏸ If $a \neq 0$, $\dfrac{a}{0}$ is _____.

Section 1.7 Multiplying and Dividing Real Numbers

> **Zero as a Divisor**
>
> **1.** The quotient of any non-zero real number and 0 is undefined. In symbols, if $a \neq 0$, $\dfrac{a}{0}$ is undefined.
>
> **2.** The quotient of 0 and any real number except 0 is 0. In symbols, if $a \neq 0$, $\dfrac{0}{a} = 0$.

Objective 4 **Evaluate expressions using real numbers**

⏺ **Work Video Exercise 16 with me.**

Evaluate.

16. $\dfrac{6 - 2(-3)}{4 - 3(-2)}$

Now press Pause ⏸ **and work Video Exercise 17. Then press Play** ▶ **and check your work.**

17. Evaluate $\dfrac{x^2 + y}{3y}$ when $x = -5$ and $y = -3$.

$$\dfrac{x^2 + y}{3y}$$

▶ Play and check.

Objective 5 **Solve applications that involve multiplication or division of real numbers**

⏺ **Work Video Exercise 18 with me.**

18. A football team lost four yards on each of three consecutive plays. Represent the total loss as a product of signed numbers and find the total loss.

Total loss =

Section 1.8 Properties of Real Numbers

Complete the outline as you view Video Lecture 1.8. Pause ⏸ the video as needed to fill in the blanks. Then press Play ▶ to continue. Also, circle your answer to each numbered exercise.

Objective 1 Use the commutative and associative properties

Commutative Properties
Addition: $a+b=b+a$
Multiplication: $a \cdot b = b \cdot a$

⏸ Commutative properties have to do with_____.

▶ **Work Video Exercise 1 with me.**

Use the commutative property to complete:

1. $x+16 = $ ____ + ____

Now press Pause ⏸ and work Video Exercise 2. Then press Play ▶ and check your work.

Use the commutative property to complete:

2. $xy = $

▶ Play and check.

▶ **Work Video Exercise 3 with me.**

Determine if the equation is true.

3. $(2+3)+4 \overset{?}{=} 2+(3+4)$ Notice the order is the same but the grouping has changed.

⏸ **Associative Properties:** Grouping does not matter when _____ or when _____.

Associative Properties
Addition: $(a+b)+c = a+(b+c)$
Multiplication: $(a \cdot b) \cdot c = a \cdot (b \cdot c)$

27

Section 1.8 Properties of Real Numbers

▶ **Work Video Exercises 4 and 5 with me.**

Use the Associative property to complete:

4. $(xy) \cdot z =$ 5. $(a+b)+c =$

Now press Pause ⏸ **and work Video Exercises 6 and 7. Then press Play** ▶ **and check your work.**

Use the associative property to complete. Simplify after you regroup.

6. $8+(9+b)$ 7. $4(6y)$

▶ Play and check.   ▶ Play and check.

⟨**Objective 2**⟩ **Use the distributive property**

Distributive Property of Multiplication over Addition

$$a(b+c) = ab + ac$$

▶ **Work Video Exercises 8 and 9 with me.**

Use the distributive property to write without parentheses. Then simplify if possible.

8. $3(6+x)$ 9. $-(r-3-7p)$

Now press Pause ⏸ **and work Video Exercise 10. Then press Play** ▶ **and check your work.**

Use the distributive property to write without parentheses. Then simplify if possible.

10. $-9(4x+8)+2$

▶ Play and check.

Section 1.8 Properties of Real Numbers

▶ Work Video Exercise 11 with me.

Use the distributive property to write the sum as a product.

11. $11x + 11y$

⬭**Objective 3**⬭ **Use the identity and inverse properties**

> ⏸ **Identities for Addition and Multiplication**
> 0 is the identity element for _____. $a + 0 = a$ and $0 + a = a$
> 1 is the identity element for _____. $a \cdot 1 = a$ and $1 \cdot a = a$

> **Additive or Multiplicative Inverses**
> The numbers a and $-a$ are additive inverses or opposites of each other because their sum is 0; that is, $a + (-a) = 0$.
>
> The numbers b and $\dfrac{1}{b}$ (for $b \neq 0$) are reciprocals or multiplicative inverses of each other because their product is 1; that is, $b \cdot \dfrac{1}{b} = 1$.

▶ Work Video Exercises 12 and 13 with me.

Name the property illustrated by each true statement.

12. $1 \cdot 9 = 9$

13. $6 \cdot \dfrac{1}{6} = 1$

Now press Pause ⏸ and work Video Exercises 14 and 15. Then press Play ▶ and check your work.

Study the statement and name the property that allows us to say the statement is true.

14. $0 + 6 = 6$

15. $(11 + r) + 8 = (r + 11) + 8$

▶ Play and check.

▶ Play and check.

29

Section 1.8 Properties of Real Numbers

Section 2.1 Simplifying Algebraic Expressions

Complete the outline as you view Video Lecture 2.1. Pause ⏸ the video as needed to fill in the blanks. Then press Play ▶ to continue. Also, circle your answer to each numbered exercise.

Objective 1 **Identify terms, like terms, and unlike terms**

⏸ A _____ is a number or the product of a number and variables raised to powers.

⏸ The _____ _____ of a term is the numerical factor of the term.

▶ **Work Video Exercise 1 with me.**

Identify the numerical coefficient.

1. $-3y$

Now press Pause ⏸ and work Video Exercises 2 and 3. Then press Play ▶ and check your work.

Identify the numerical coefficient.

2. $22z^4$ 3. y

▶ Play and check.   ▶ Play and check.

▶ **Work Video Exercises 4 and 5 with me.**

Identify the numerical coefficient.

4. $-x$ 5. $\dfrac{x}{7}$

⏸ Terms with the same variables raised to exactly the same powers are called_____ _____.

⏸ Terms that aren't like terms are called _____ _____.

Section 2.1 Simplifying Algebraic Expressions

⬤ **Work Video Exercises 6 and 7 with me.**

Identify whether the terms are like or unlike terms.

6. $5y, -y$ 7. $-2x^2y, 6xy$

(**Objective 2**) **Combine like terms**

⬤ **Work Video Exercise 8 with me.**

Simplify.

8. $3x + 2x$

⏸ Simplifying the sum or difference of like terms is called _____ _____ _____.

Now press Pause ⏸ **and work Video Exercises 9 and 10. Then press Play** ▶ **and check your work.**

Simplify.

9. $8x^3 + x^3 - 11x^3$

▶ Play and check.

10. $6x + 0.5 - 4.3x - 0.4x + 3$

▶ Play and check.

Combining Like Terms
To combine like terms, add the numerical coefficients and multiply the result by the common variable factors.

Section 2.1 Simplifying Algebraic Expressions

Objective 3 **Use the distributive property to remove parentheses**

⏵ **Work Video Exercise 11 with me.**

Use the distributive property to simplify.

11. $5(x+2)-(3x-4)$

Now press Pause ⏸ **and work Video Exercise 12. Then press Play** ⏵ **and check your work.**

⏵ 12. Subtract $(5m-6)$ from $(m-9)$.

⏵ Play and check.

Objective 4 **Write word phrases as algebraic expressions**

⏵ **Work Video Exercises 13 and 14 with me.**

Translate each phrase to an algebraic expression.

13. Twice a number, decreased by four

14. The sum of 5 times a number and -2, plus 7 times a number

Section 2.1 Simplifying Algebraic Expressions

Section 2.2 The Addition Property of Equality

Complete the outline as you view Video Lecture 2.2. Pause ⏸ the video as needed to fill in the blanks. Then press Play ▶ to continue. Also, circle your answer to each numbered exercise.

Objective 1 **Define linear equations and use the addition property of equality to solve linear equations**

Equation: expression = expression

⏸ A _____ is a value for the variable that makes an equation a true statement.

⏸ _____ _____ is the process of finding a value for the variable that makes the equation a true statement.

Linear Equation in One Variable
A linear equation in one variable can be written in the form $ax + b = c$ where a, b, and c are real numbers and $a \neq 0$.

⏸ _____ _____ have the same solution.

Addition Property of Equality
If a, b, and c are real numbers, then $a = b$ and $a + c = b + c$ are equivalent equations.

▶ **Work Video Exercises 1 and 2 with me.**

Solve each equation.

1. $x - 2 = -4$

2. $\dfrac{1}{3} = x + \dfrac{2}{3}$

Check: Check:

Now press Pause ⏸ and work Video Exercise 3. Then press Play ▶ and check your work.

Solve the equation.

3. $5b - 0.7 = 6b$

▶ Play and check.

Section 2.2 The Addition Property of Equality

Work Video Exercise 4 with me.

Solve the equation.

4. $3x - 6 = 2x + 5$

Check:

Now press Pause ⏸ **and work Video Exercise 5. Then press Play** ▶ **and check your work.**

Solve the linear equation.

5. $13x - 9 + 2x - 5 = 12x - 1 + 2x$

▶ Play and check.

▶ **Work Video Exercise 6 with me.**

Solve the equation.

6. $15 - (6 - 7k) = 2 + 6k$

Objective 2 **Write word phrases as algebraic expressions**

▶ **Work Video Exercise 7 with me.**

7. Two numbers have a sum of 20. If one number is p, express the other number in terms of p.

Now press Pause ⏸ **and work Video Exercise 8. Then press Play** ▶ **and check your work.**

8. The area of the Sahara Desert in Africa is 7 times the area of the Gobi Desert in Asia. If the area of the Gobi Desert is x square miles, express the area of the Sahara Desert as an algebraic expression in x.

▶ Play and check.

Section 2.3 The Multiplication Property of Equality

Complete the outline as you view Video Lecture 2.3. Pause ⏸ the video as needed to fill in the blanks. Then press Play ▶ to continue. Also, circle your answer to each numbered exercise.

Objective 1 Use the multiplication property of equality to solve linear equations

---

Multiplication Property of Equality
If a, b, and c are real numbers and $c \neq 0$, then $a = b$ and $ac = bc$ are equivalent equations.

▶ **Work Video Exercises 1 and 2 with me.**

Solve each equation.

1. $-5x = -20$

2. $\dfrac{2}{3}x = -8$

⏸ **Multiplication Property**
We can _____ or _____ both sides of an equation by the same non-zero number and have an equivalent equation.

Now press Pause ⏸ and work Video Exercise 3. Then press Play ▶ and check your work.

Solve the equation for d.

3. $\dfrac{d}{15} = 2$

▶ Play and check.

Objective 2 Use both the addition and multiplication properties of equality to solve linear equations

▶ **Work Video Exercises 4, 5, and 6 with me.**

Solve each equation.

4. $6a + 3 = 3$

Section 2.3 The Multiplication Property of Equality

5. $8x + 20 = 6x + 18$

6. $-10z - 0.5 = -20z + 1.6$

Now press Pause ⏸ **and work Video Exercise 7. Then press Play** ▶ **and check your work.**

Solve the linear equation.

7. $9(3x+1) = 4x - 5x$

▶ Play and check.

(**Objective 3**) **Write word phrases as algebraic expressions**

▶ **Work Video Exercise 8 with me.**

8. If x is the first of four consecutive integers, express the sum of the first integer and the third integer as an algebraic expression containing the variable x.

$$x =$$
$$x + 1 =$$
$$x + 2 =$$
$$x + 3 =$$

Consecutive Integers: x, $x+1$, $x+2$, $x+3$

Complete the outline as you view Video Lecture 2.4. Pause ⏸ the video as needed to fill in the blanks. Then press Play ▶ to continue. Also, circle your answer to each numbered exercise.

Objective 1 **Apply a general strategy for solving a linear equation**

▶ **Work Video Exercise 1 with me.**

Solve the equation.

1. $5(2x-1)-2(3x)=1$

⏸ **Solving Linear Equations in One Variable**
Step 1: Multiply on both sides by the _____ to clear the equation of fractions if they occur.
Step 2: Use the _____ _____ to remove parentheses if they occur.
Step 3: Simplify each side of the equation by combining _____ _____.
Step 4: Get all variable terms on one side and all numbers on the other side by using the _____ _____ of _____.
Step 5: Get the variable alone by using the _____ _____ of _____.
Step 6: Check the solution by substituting it into the original equation.

Objective 2 **Solve equations containing fractions**

Now press Pause ⏸ and work Video Exercise 2. Then press Play ▶ and check your answer.

Solve the linear equation. Fill in the steps on the right as you work through the problem.

2. $\dfrac{x}{2}-1=\dfrac{x}{5}+2$

⏸ **Step 1:** Multiply on both sides by the _____.
Step 2: Use the _____ _____ to remove the parentheses.
Step 3: Simplify.
Step 4: Get all variable terms on one side and numbers on the other side by using the _____.
Step 5: Get the variable alone by using the _____.

▶ Play and check.

Section 2.4 Solving Linear Equations

Objective 3 Solve equations containing decimals

▶ **Work Video Exercise 3 with me.**

Solve the equation.

3. $0.50x + 0.15(70) = 35.5$

Objective 4 Recognize identities and equations with no solution

> ❙❙ Some equations have _____ _____ .
> Some equations have all _____ _____ as solutions.

▶ **Work Video Exercises 4 and 5 with me.**

Solve each equation.

4. $5x - 5 = 2(x + 1) + 3x - 7$

5. $2(x + 3) - 5 = 5x - 3(1 + x)$

> ❙❙ If variable terms subtract out, giving us a <u>true</u> statement, then the equation has _____
> _____ _____ as a solution.

> ❙❙ If variable terms subtract out, giving a <u>false</u> statement, then the equation has _____
> _____ .

Complete the outline as you view Video Lecture 2.5. Pause ⏸ the video as needed to fill in the blanks. Then press Play ▶ to continue. Also, circle your answer to each numbered exercise.

Objective 1 **Solve problems involving direct translations**

Translate the problem into an equation and solve.

▶ **Work with me.**

1. Twice the difference of a number and 8 is equal to three times the sum of the number and 3. Find the number.

   x = The number

⏸ **General Strategy for Problem Solving**

   Step 1: _____ the problem. During this step, become comfortable with the problem.
   Some ways of doing this are:
   Read and reread the problem.
   Choose a variable to represent the unknown.
   Construct a drawing, whenever possible.
   Propose a solution and check. Pay careful attention to how you check your proposed solution. This will help when writing an equation to model the problem.
   Step 2: _____ the problem into an equation.
   Step 3: _____ the equation.
   Step 4: _____ the results: Check the proposed solution in the stated problem and state your conclusion.

Objective 2 **Solve problems involving relationships among unknown quantities**

Use the problem solving steps above to solve the following applications.

▶ **Work with me.**

2. The area of the Sahara Desert is 7 times the area of the Gobi Desert. If the sum of their areas is 4,000,000 square miles, find the area of each desert.

   x = area of Gobi Desert
   $7x$ = area of the Sahara Desert

Section 2.5 An introduction to Problem Solving

⊙ **Work with me.**

3. Two angles are supplementary if their sum is $180°$. The larger angle measures eight degrees more than three times the measure of a smaller angle. If x represents the measure of the smaller angle and these two angles are supplementary, find the measure of each angle.

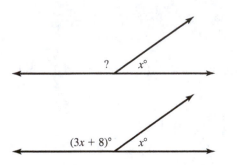

⬭ **Objective 3** Solve problems involving consecutive integers

Solve.

⊙ **Work with me.**

4. The measures of the angles of a triangle are 3 consecutive even integers. Find the measure of each angle.

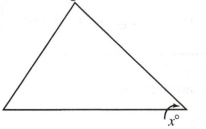

$x =$ an even integer
_____ = next consecutive even integer
_____ = third consecutive even integer

$$x + (x+2) + (x+4) = 180$$

Section 2.6 Formulas and Problem Solving

Complete the outline as you view Video Lecture 2.6. Pause ⏸ the video as needed to fill in the blanks. Then press Play ▶ to continue. Also, circle your answer to each numbered exercise.

Objective 1 **Use formulas to solve problems**

⏸ An equation that describes a known relationship among quantities is called a _____.

▶ **Work with me.**
1. Convert Nome, Alaska's 14° F high temperature to Celsius.

$$C = \frac{5}{9}(F - 32)$$

▶ **Work with me.**

2. The Cat is a high-speed catamaran auto ferry that operates between Bar Harbor, Maine and Yarmouth, Nova Scotia. The Cat can make the 138 mile trip in about 2 ½ hours. Find the Cat's speed for this trip.

Distance = rate · time
$$d = \quad r \quad \cdot \quad t$$

⏸ **Pause and work.**

3. An architect designs a rectangular flower garden such that the width is exactly two-thirds of the length. If 260 feet of antique picket fencing are to be used to enclose the garden, find the dimensions of the garden.

⏸ _____ means distance around a figure.

x

x

▶ Play and check.

Section 2.6 Formulas and Problem Solving

Objective 2 Solve a formula or equation for one its variables

---

Solving Equations for a Specified Variable
Step 1: Multiply on both sides to clear the equation of fractions if they occur.
Step 2: Use the _____ _____ to remove parentheses if they occur.
Step 3: Simplify each side of the equation by combining _____ _____.
Step 4: Get all terms containing the specified variable on one side and all other terms on the other side by using the _____ _____ of _____.
Step 5: Get the specified variable alone by using the _____ _____ of _____.

Work with me.

4.   Solve $V = \ell wh$ for w.

Work with me.

5.   Solve $S = 2\pi rh + 2\pi r^2$ for h.

Section 2.7 Percent and Mixture Problem Solving

Complete the outline as you view Video Lecture 2.7. Pause ⏸ the video as needed to fill in the blanks. Then press Play ▶ to continue. Also, circle your answer to each numbered exercise.

Objective 1 **Solve percent equations**

⏸ **Things to Know When Working with Percents**
100% of a number is that number.
50% of a number is _____ that number.
25% of a number is _____ of that number.
To find 10% of a number, move the decimal point of the number 1 place to the _____.

⏸ Review the Steps for Problem Solving:

General Strategy for Problem Solving
Step 1: _____ the problem.
Step 2: _____ the problem into an equation.
Step 3: _____ the equation.
Step 4: _____ the results.

▶ **Work with me.**

1.   Find 23% of 20.

⏸ **Pause and work.**

2.   The number 45 is 25% of what number?

▶ Play and check.

⏸ "of" means _____
   "is" means _____

Section 2.7 Percent and Mixture Problem Solving

Objective 2 **Solve discount and mark-up problems**

__________ = percent · original price __________ __________ = original price + mark-up

Solve.

Work with me.

3. A birthday celebration meal is $40.50 including tax. Find the total cost if a 15% tip is added to the cost.

   Mark-up =

   New price =

Objective 3 **Solve percent increase and percent decrease problems**

Solve.

Work with me.

4. Due to arid conditions in lettuce growing areas at the end of 2014, the cost of growing and shipping iceberg lettuce to stores was about $1.19. In certain areas, consumers purchased lettuce for $1.49 a head. Find percent of increase and round to the nearest tenth of a percent.

   increase = higher cost − lower cost

   increase = what % of lower cost

🔘 **Work with me.**

5. Find last year's salary if after a 4% pay raise, this year's salary is $ 44,200.

$$x = \text{old salary}$$
$$\text{old salary} + \text{raise} = \text{new salary}$$

(Objective 4) **Solve mixture problems**

Solve.

🔘 **Work with me.**

6. How much of an alloy that is 20% copper should be mixed with 200 ounces of an alloy that is 50% copper in order to get an alloy that is 30% copper?

Organize the information in a table.

Alloy	Ounces · Copper Strength = Amount of Copper

Translate to an equation and solve.

Complete the outline as you view Video Lecture 2.8. Pause 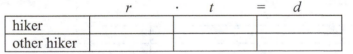 the video as you fill in the blanks. Then press Play to continue. Also, circle your answer to each numbered exercise.

Objective 1 Solve problems involving distance

Solve.

Work with me.

1. How long will it take a bus traveling at 60 mph to overtake a car traveling at 40 mph if the car had a 1.5 hour head start?

Organize the information in a chart.

	r	$\cdot$	t	$=$	d
bus			
car			

Translate the information into an equation and solve.

Work with me.

2. Two hikers are 11 miles apart and walking toward each other. They meet in two hours. Find the rate of each hiker if one hiker walks 1.1 mph faster than the other.

Organize the information in a chart.

	r	$\cdot$	t	$=$	d
hiker			
other hiker			

Translate the information into equation and solve.

Objective 2 Solve problems involving money

Understand the difference between the _____ of bills vs. the _____ of bills.

Section 2.8 Further Problem Solving

Solve.

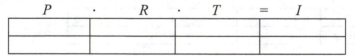 **Work with me.**

3. Part of the proceeds from a garage sale was $280 worth of $5 and $10 bills. If there were 20 more $5 bills than $10 bills, find the number of each denomination.

$$x =$$
$$x + 20 =$$

⟨Objective 3⟩ **Solve problems involving interest**

⓫ Interest = _____ • _____ • _____

Solve.

Work with me.

4. How can $54,000 be invested, part at 8% annual simple interest and the remainder at 10% annual simple interest, so that the annual interest earned by the two accounts is equal?

Organize the information in a chart.

P	•	R	•	T	=	I

Translate the information into equation and solve.

Section 2.9 Solving Linear Inequalities

Complete the outline as you view Video Lecture 2.9. Pause 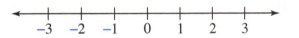 the video as you fill in the blanks. Then press Play ▶ to continue. Also, circle your answer to each numbered exercise.

⬭ Objective 1 ⬭ **Define a linear inequality in one variable, graph solution sets on a number line, and use interval notation**

> ⏸ **Review of Inequality Symbols**
>   < means _____ .
>   > means _____ .
>   ≤ means _____ .
>   ≥ means _____ .

Graph the inequality on the number line and write the answer in interval notation.

▶ **Work with me.**

1. $x \le -1$

```
  ←——+——+——+——+——+——+——+——→
     −3  −2  −1   0   1   2   3
```

> ⏸ **≤ or ≥ :** Use] or [instead of _____ .

> ⏸ **Interval notation:** Write down shading from _____ to _____ .

> Place a parenthesis about $-\infty$ or ∞ .

Graph the inequality on the number line and write the answer in interval notation.

⏸ **Pause and work.**

2. $x > 3$

```
  ←——+——+——+——+——+——+——+——+——+——→
     −4  −3  −2  −1   0   1   2   3   4
```

> ⏸ **< or > :** Use) or (instead of _____ .

Section 2.9 Solving Linear Inequalities

Linear Inequality in One Variable
A linear inequality in one very variable is an inequality that can be written in the form $ax + b < c$ where a, b, and c are real numbers and $a \neq 0$.

(Objective 2) **Solve linear inequalities**

Addition Property of Inequality
If a, b, and c are real numbers, then $a < b$ and $a + c < b + c$ are equivalent inequalities.

Solve the linear inequality for x. Graph the solution set and write it in interval notation.

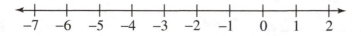 **Work with me.**

3. $x - 2 \geq -7$

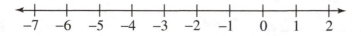

    Interval notation:

Be careful: If we multiply or divide both sides of an inequality by a negative number, _____ the direction of the inequality symbol.

Multiplication Property of Inequality
1. If a, b, and c are real numbers, and c is **positive**, then $a < b$ and $ac < bc$ are equivalent inequalities.
2. If a, b, and c are real numbers, and c is **negative,** then $a < b$ and $ac > bc$ are equivalent inequalities.

Solve each linear inequality. Graph the solution set and write it in interval notation.

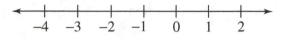 **Work with me.**

4. $-8x \leq 16$

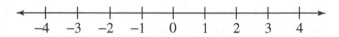

Interval notation:

⏸ **Pause and work.**

5. $2x < -4$

Interval notation:

▶ Play and check.

⏸ **Solving Linear Inequalities in One Variable**
Step 1: Clear the inequality of fractions by multiplying both sides of the inequality by the _____ _____ _____ (____) of all fractions in the inequality.
Step 2: Remove grouping symbols such as parentheses by using the_____ _____.
Step 3: _____ each side of the inequality by combining like terms.
Step 4: Write the inequality with variable terms on one side and numbers on the other side by using the _____ _____ of_____.
Step 5: Get the variable alone by using the_____ _____ of _____.

Solve the linear inequality. Graph the solution set and write it in interval notation.

⏸ **Pause and work.**

6. $-2(x-4)-3x < -(4x+1)+2x$

Interval notation:

▶ Play and check.

Section 2.9 Solving Linear Inequalities

Objective 3 **Solve compound inequalities**

⏸ Inequalities containing two inequality symbols are called _____ _____.

Graph the compound inequality. Then write the solution in interval notation.

▶ **Work with me.**

7. $0 \le y < 2$

Interval notation:

Solve the compound inequality. Graph the solution set and write it in interval notation.

▶ **Work with me.**

8. $-6 < 3(x - 2) \le 8$

Interval notation:

Objective 4 **Solve inequality applications**

▶ **Work with me.**

9. Find the values for x so that the perimeter of this rectangle is no greater than 100 centimeters.

x cm

15 cm

⏸ Perimeter is the _____ around.

Solve.

Section 3.1 Reading Graphs and the Rectangular Coordinate System

Complete the outline as you view Video Lecture 3.1. Pause ⏸ the video as needed to fill in the blanks. Then press Play ▶ to continue. Also, circle your answer to each numbered exercise.

(Objective 1)   **Read bar and line graphs**

⏸ A _____ _____ consists of a series of vertical or horizontal bars.

Study the bar graph for the top 10 tourist destinations and the number of tourists that visit each country per year. Use this bar graph to answer Video Exercises 1, 2, and 3.

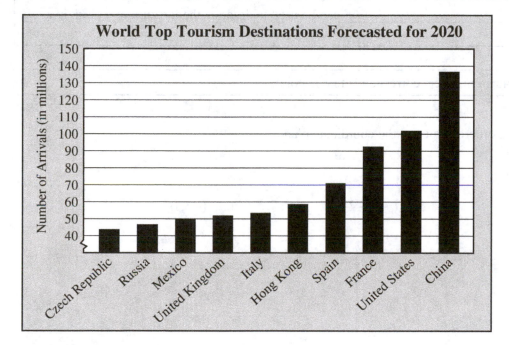

▶ **Work with me.**

1.   Which country in the graph is the most popular tourist destination?

▶ **Work with me.**

2.   Which countries in the graph have more than 70 million tourists per year?

Section 3.1 Reading Graphs and the Rectangular Coordinate System

⏸ **Pause and work.**

3.   Predict the number of tourists per year in Italy.

▶ Play and check.

(Objective 2) **Define the rectangular or coordinate system and plot ordered pairs of numbers**

⏸ The _____ _____ _____ can be used to describe the location of a point in a plane.

A single ordered pair of numbers corresponds to one point.

Graph the following ordered pairs on the coordinate plane.

▶ **Work with me.**

4. $(1, 5)$ 5. $(-5, -2)$ 6. $(-3, 0)$

⏸ **Pause and work.**

7. $(0, -1)$ 8. $(2, -4)$ 9. $\left(-1, 4\frac{1}{2}\right)$

▶ Play and check.

Section 3.1 Reading Graphs and the Rectangular Coordinate System

---

ⅠⅠ An _____ _____ of numbers is of the form (x, y).

ⅠⅠ The first coordinate tells us to move _____ or _____. The second coordinate tells us
to move _____ or _____.

ⅠⅠ When the y-value is 0, the point lies on the _____.
When the x-value is 0, the point lies on the _____.

(Objective 3) **Graph paired data to create a scatter diagram**

▶ **Work with me.**

10. Study the table showing selected cities, their distance from the equator (in miles), and the
average snowfall (in inches). Write this paired data as a set of ordered pairs of the form
(distance from the equator, average annual snowfall).

City	Distance from Equator (in miles)	Average Annual Snowfall (in inches)
Atlanta, GA	2313	2
Baltimore, MD	2711	21
Detroit, MI	2920	42
Miami, FL	1783	0

Ordered pairs:

Section 3.1 Reading Graphs and the Rectangular Coordinate System

11. Plot the ordered pairs to create a scatter diagram.

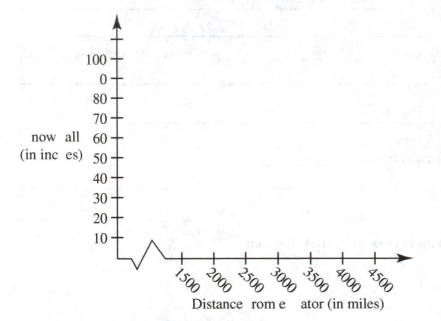

now all
(in inc es)

Distance rom e   ator (in miles)

Objective 4 **Determine whether an ordered pair is a solution of an equation in two variables**

A _____ of an equation in x and y consists of a value for x and a value for y so that a <u>true</u> <u>statement</u> results.

Determine if the given ordered pair is a solution of an equation in two variables.

Work with me.

12. Given the equation $2x + y = 7$, are the following points solutions?

$(3,1)$:

$(7,0)$:

Section 3.1 Reading Graphs and the Rectangular Coordinate System

⏸ Pause and work.

$(0,7)$:

▶ Play and check.

⏸	
_____ statement: ordered pair <u>is</u> a solution.	
_____ statement: ordered pair <u>is not</u> a solution.	

Objective 5 **Find the missing coordinate of an ordered pair solution, given one coordinate of the pair**

Find the missing coordinate of the ordered pair.

▶ Work with me.

13. Given the equation $x - 4y = 4$, complete the ordered pairs to make a solution to the equation.

$(\quad, -2)$:

▶ Pause and work.

$(4, \quad)$:

▶ Play and check.

Section 3.1 Reading Graphs and the Rectangular Coordinate System

Complete the outline as you view Video Lecture 3.2. Pause ⏸ the video as needed to fill in the blanks. Then press Play ▶ to continue. Also, circle your answer to each numbered exercise.

Objective 1 **Identify linear equations**

> ⏸ **Linear Equations in Two Variables**
> A linear equation in two variables is an equation that can be written in the form
> _____ where A, B, and C are real numbers and A and B are not both 0. The graph of a linear equation in two variables is a straight line.

> ⏸ The form $Ax + By = C$ is called _____ _____.

Determine whether each equation is a linear equation in two variables.

▶ **Work with me.**

1. $x - 1.5y = -1.6$

⏸ **Pause and work.**

2. $y = -2x$

▶ Play and check.

▶ **Work with me.**

3. $x + y^2 = 9$

▶ **Work with me.**

4. $x = 5$

Objective 2 **Graph a linear equation by finding and plotting ordered pair solutions**

Graph the linear equation.

▶ **Work with me.**

5. $x - 2y = 6$

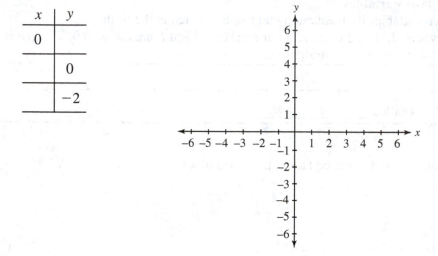

x	y
0	
	0
	−2

⏸ Every _____ corresponds to an _____ _____ that is a _____ of the linear equation.

Graph each linear equation.

⏸ **Pause and work.**

6. $x = -3y$

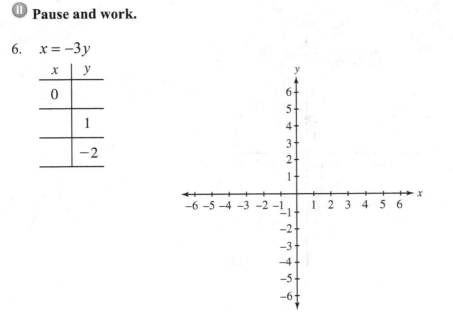

x	y
0	
	1
	-2

▶ Play and check.

Section 3.2 Graphing Linear Equations

Work with me.

7. $y = \dfrac{1}{2}x + 2$

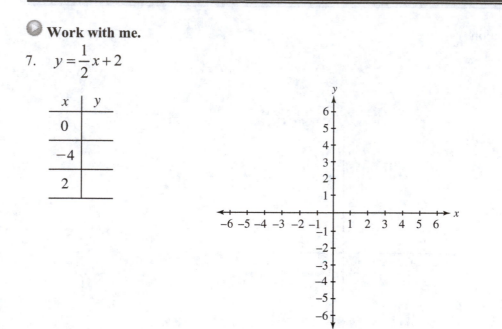

x	y
0	
−4	
2	

Complete the outline as you view Video Lecture 3.3. Pause the video as needed to fill in the blanks. Then press Play 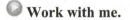 to continue. Also, circle your answer to each numbered exercise.

(Objective 1) **Identify intercepts of a graph**

Identify the intercepts.

▶ **Work with me.**

1.

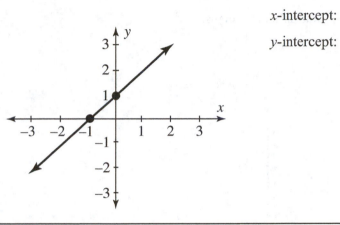

x-intercept:

y-intercept:

⏸ A point where a graph crosses the *y*-axis is called a _____.
A point where a graph crosses the *x*-axis is called an _____.

Identify the intercepts.

⏸ **Pause and work.**

2.

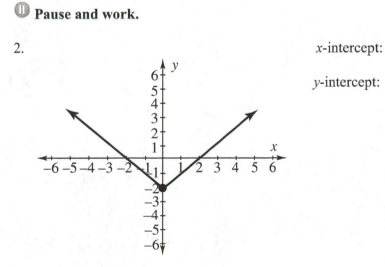

x-intercept:

y-intercept:

▶ Play and check.

Section 3.3 Intercepts

⓫ *x*-intercepts have *y*-values of _____ .
    y-intercepts have *x*-values of _____ .

Objective 2 **Graph a linear equation by finding and plotting intercepts**

To find the *x*-intercepts, let *y* equal 0 and solve for *x*.
To find the *y*-intercept, let *x* equal 0 and solve for *y*.

Graph each linear equation by finding and plotting its intercepts.

▶ **Work with me.**

3. $-x + 2y = 6$

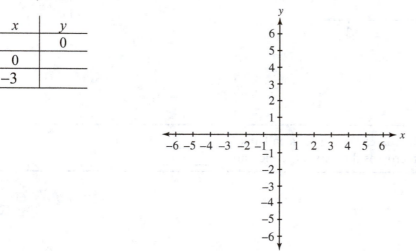

x	y
	0
0	
−3	

Pause and work.

4. $y = -2x$

x	y
0	
	0
1	
-1	

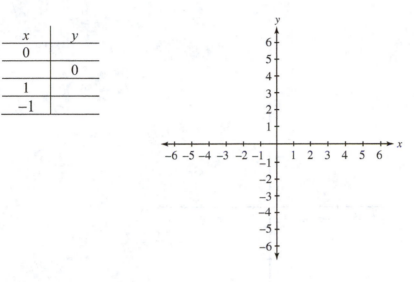

▶ Play and check.

 If an ordered pair solution is (0,0), the x-intercept is (___ , ___) and the y-intercept is (___ , ___).

Objective 3 Identify and graph vertical and horizontal lines

Graph the following linear equation.

Work with me.

5. $y = 5$

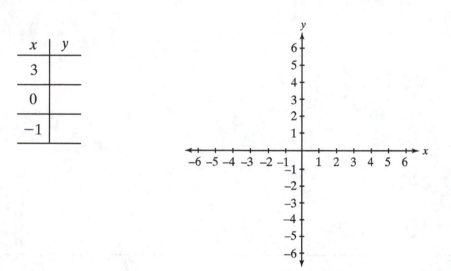

x	y
3	
0	
−1	

Horizontal lines

The graph of $y = c$, where c is a real number, is a horizontal line with y-intercept $(0, c)$.

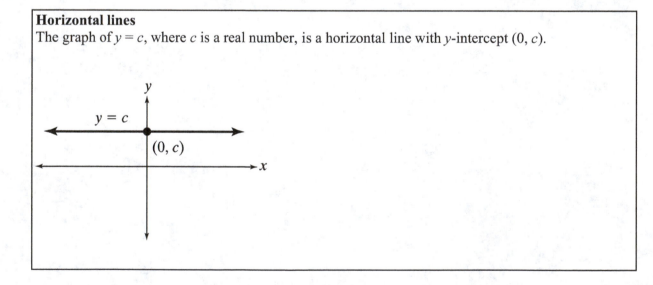

Graph the linear equation.

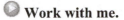 **Work with me.**

6. $x = -3$

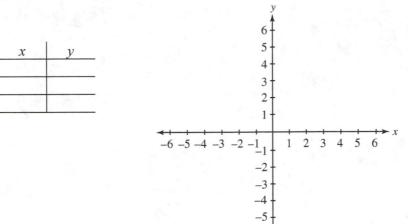

x	y

Vertical lines
The graph of $x = c$, where c is a real number, is a vertical line with x-intercept $(c, 0)$.

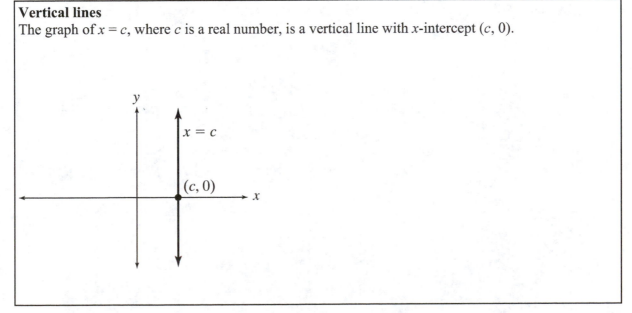

Section 3.4 Slope and Rate of Change

Complete the outline as you view Video Lecture for 3.4. Pause ⏸ the video as needed to fill in the blanks. Then press Play ▶ to continue. Also, circle your answer to each numbered exercise.

Objective 1 **Find the slope of a line given two points of the line**

⏸ $\underline{\hspace{2cm}} = \dfrac{\text{vertical change}}{\text{horizontal change}}$

⏸ $y_2 - y_1$ vertical change or $\underline{\hspace{3cm}}$

$x_2 - x_1$ horizontal change or $\underline{\hspace{2.5cm}}$

Slope of a Line
The slope m of the line containing points (x_1, y_1) and (x_2, y_2) is given by:

$$m = \frac{\text{rise}}{\text{run}} = \frac{\text{change in } y}{\text{change in } x} = \frac{y_2 - y_1}{x_2 - x_1} \text{ as long as } x_2 \neq x_1.$$

Find the slope of the line.

▶ **Work with me.**

1.

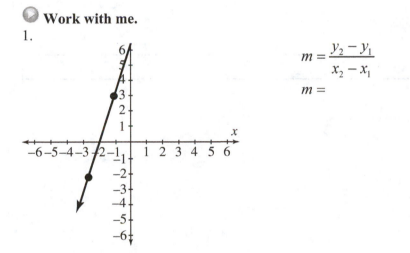

$$m = \frac{y_2 - y_1}{x_2 - x_1}$$

$$m =$$

Whatever y-value is the first in the numerator, the corresponding x-value <u>must</u> <u>also</u> <u>be</u> <u>first</u>.

⏸ **Positive slope:** line moves upward from$\underline{\hspace{2cm}}$ to $\underline{\hspace{2cm}}$.

Section 3.4 Slope and Rate of Change

Find the slope of a line that passes through the given points.

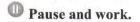 **Pause and work.**

2. $(-1, 5), (6, -2)$

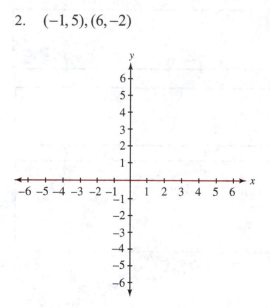

▶ Play and check.

Negative slope: line moves downward from _____ to _____.

Find the slope of the line through the given points.

▶ **Work with me.**

3. $(-4, 3), (-4, 5)$

Undefined slope: _____ _____.

Section 3.4 Slope and Rate of Change

Find the slope of the line that passes through the given points. Then graph the line through the points.

❚❚ Pause and work.

4. $(5, 1) (-2, 1)$

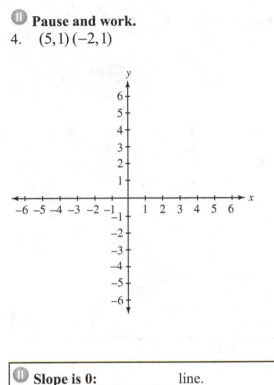

❚❚ Slope is 0: _____ line.

Objective 2 **Find the slope of a line given its equation**

Slope Intercept Form
When a linear equation in two variables is written in slope-intercept form, $y = mx + b$, m is the slope of the line and $(0, b)$ is the y-intercept of the line.

▶ **Work with me.**

5. Find the slope and the y-intercept of the line whose equation is $2x + y = 7$.

❚❚ When an equation is solved for y, the coefficient of x is the_____.

73

Section 3.4 Slope and Rate of Change

⏸ **Pause and work.**

6. Find the slope and y-intercept of the graph of the equation $2x - 3y = 10$.

▶ Play and check.

(**Objective 3**) **Find the slopes of horizontal and vertical lines**

Find the slope of the line.

▶ **Work with me.**

7. $x = 1$

⏸ **Vertical line:** _____ slope

▶ **Work with me.**

8. $y = -3$

⏸ **Horizontal line:** $m =$ _____

⏸ **Review of slopes**

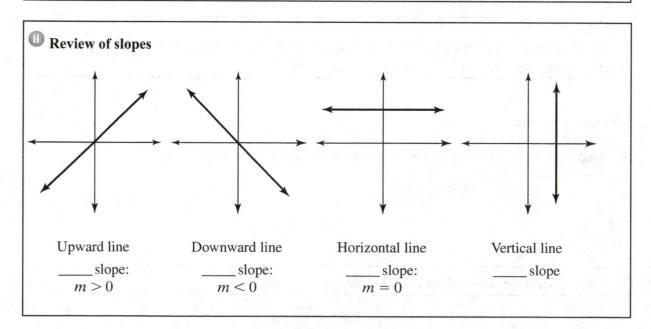

Upward line	Downward line	Horizontal line	Vertical line
_____ slope:	_____ slope:	_____ slope:	_____ slope
$m > 0$	$m < 0$	$m = 0$	

Objective 4 **Compare the slopes of parallel and perpendicular lines**

Parallel Lines
Nonvertical parallel lines have the same slope and different *y*-intercepts.

Perpendicular Lines
If the product of the slopes of the two lines is -1, then the lines are perpendicular. Two non-vertical lines are perpendicular if the slope of one is the negative reciprocal of the slope of the other.

Determine whether each pair of lines is parallel, perpendicular, or neither.

▶ **Work with me.**

9. $y = \dfrac{2}{9}x + 3$ $m =$

     $y = \dfrac{-2}{9}x$ $m =$

By comparing the slopes, we see that these lines are neither _____ nor

_____ .

⏸ **Pause and work.**

10. $10 + 3x = 5y$
       $5x + 3y = 1$

The two lines are _____ .

▶ Play and check.

Objective 5 **Interpret slope as a rate of change**

▶ **Work with me.**

11.   The graph in the video shows the cost of owning and operating a compact car. Find and interpret the slope of the line.

It costs _____ per _____ mile to own and operate.

Section 3.4 Slope and Rate of Change

Section 3.5 Equations of Lines

Complete the outline as you view Video Lecture 3.5. Pause ⏸ the video as needed to fill in the blanks. Then press Play ▶ to continue. Also, circle your answer to each numbered exercise.

Objective 1 Use the slope-intercept form to graph a linear equation

Slope-Intercept Form
When a linear equation in two variables is written in slope-intercept form,

$$y = mx + b$$

↑   ↑

Slope $(0, b)$ y-intercept

m is the slope of the line and $(0, b)$ is the y-intercept of the line.

Use the slope-intercept form to graph the equation.

▶ **Work with me.**

1. $y = -5x$

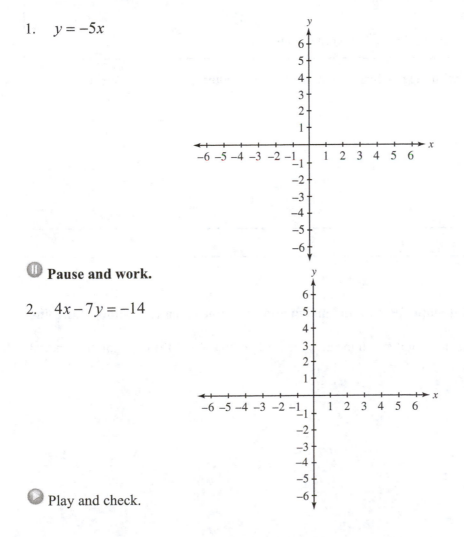

⏸ **Pause and work.**

2. $4x - 7y = -14$

▶ Play and check.

Section 3.5 Equations of Lines

$\boxed{\text{Objective 2}}$ **Use the slope-intercept form to write an equation of a line**

Write an equation of the line with slope m and y-intercept $(0, b)$

▶ **Work with me.**

3. $m = -4, \ b = -\dfrac{1}{6}$

$\boxed{\text{Objective 3}}$ **Use the point-slope form to find an equation of a line given its slope and a point on the line**

Point-Slope Form of the Equation of a line
The point-slope form of the equation of the line is $$y - y_1 = m(x - x_1)$$ where m is the slope of the line and (x_1, y_1) is a point on the line.

Find an equation of the line with the given slope that passes through the given point. Write the equation in standard form.

▶ **Work with me.**

4. $m = -8; (-1, -5)$

⏸ Point-Slope Form: _____

$\boxed{\text{Objective 4}}$ **Use the point-slope form to find an equation of a line given two points of the line**

Write an equation of the line passing through the given pair of points. Write the equation in standard form.

▶ **Work with me.**

5. $(2, 3), (-1, -1)$

Section 3.5 Equations of Lines

Objective 5 Find the equations of vertical and horizontal lines

▶ **Work with me.**

6. Write the equation of the line parallel to $y = 5$ and passing through the point $(1, 2)$.

> ⏸ **Horizontal Line Form:**_____

⏸ **Pause and work.**

7. Write the equation of the line with undefined slope and passing through the point $\left(\dfrac{-3}{4}, 1 \right)$.

▶ Play and check.

> ⏸ **Vertical Line Form:** _____

Objective 6 Use the point-slope form to solve problems

Solve.

▶ **Work with me.**

8. A rock is dropped from the top of a 400-foot cliff. After one second, the rock is traveling 32 feet per second. After 3 seconds, the rock is traveling at 96 feet per second.

    a. Assume that the relationship between time and speed is linear and write an equation describing this relationship. Use ordered pairs of the form (time, speed).

        Use the given information to write two ordered pairs: (_____, _____) and (_____ , _____).

        Use the point-slope form to write an equation. First, find the slope.

        Next, use the slope and either one of the points to write the equation in point-slope form.

    b. Use this equation to determine the speed of the rock 4 seconds after it was dropped.

79

Section 3.5 Equations of Lines

Complete the outline as you view Video Lecture 3.6. Pause ⏸ the video as needed to fill in the blanks. Then press Play ▶ to continue. Also, circle your answer to each numbered exercise.

Objective 1 **Identify relations, domains, and ranges**

⏸ A set of ordered pairs is a _____ .

⏸ The set of x-coordinates in a relation is called the _____ of the relation.

⏸ The set of y-coordinates in a relation is called the _____ of the relation.

Find the domain and range of the relation.

▶ **Work with me.**

1. $\{(0, -2), (1, -2), (5, -2)\}$

   Domain:
   Range:

Objective 2 **Identify functions**

Function
A function is a set of ordered pairs that assigns to each x-value exactly one y-value.

Determine if each relation is also a function.

▶ **Work with me.**

2. $\{(1, 1), (2, 2), (-3, -3), (0, 0)\}$

⏸ **Pause and work.**

3. $\{(-1, 0), (-1, 6), (-1, 8)\}$

▶ Play and check.

Section 3.6 Functions

Determine if the graph is also a function.

 **Work with me.**

4.

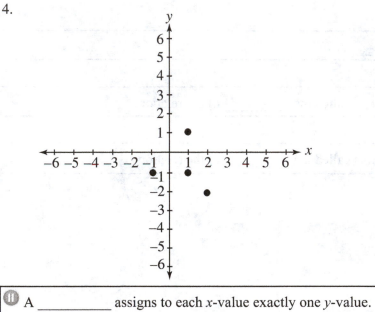

A _____ assigns to each *x*-value exactly one *y*-value.

Determine if the graph is also a function.

Work with me.

5.

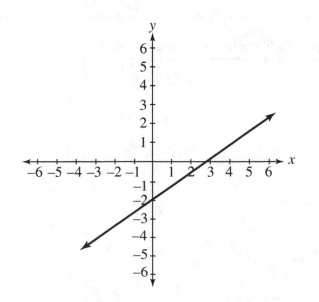

Objective 3   Use the vertical line test

Vertical Line Test
If a vertical line can be drawn so that it intersects a graph more than once, the graph is not the graph
of a function.

Use a vertical line test to determine if the graph is the graph of a function.

▶ **Work with me.**   ▶ **Work with me.**

6.   7.

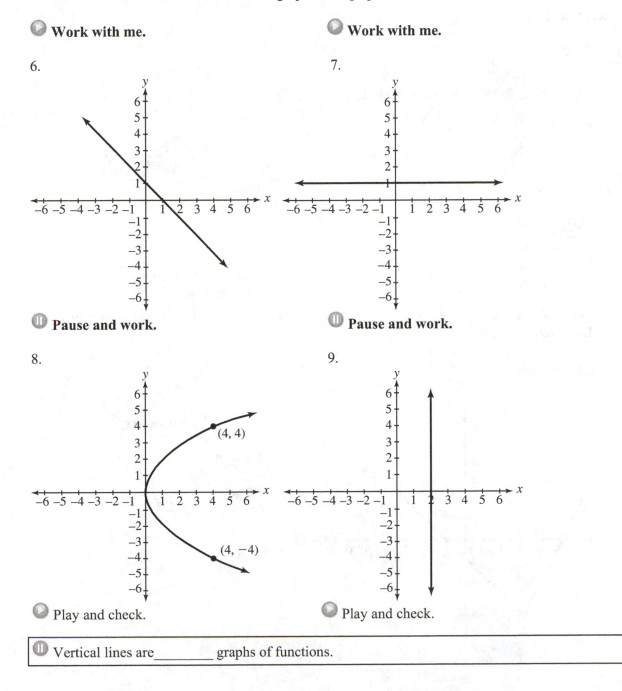

⏸ **Pause and work.**   ⏸ **Pause and work.**

8.   9.

▶ Play and check.   ▶ Play and check.

⏸ Vertical lines are_____ graphs of functions.

Objective 4   **Use function notation**

⏸ $f(x)$ does not mean multiplication. It is a special _____ _____.

Given $f(x) = x^2 + 2$, find the following. Then write the corresponding ordered pairs generated.

▶ **Work with me.**

10. $f(-2)$

   $f(-2) = $ _____ corresponds to (_____, _____)

   $f(0)$

⏸ **Pause and work.**

   $f(3)$

▶ **Play and check.**

Use the vertical line test to determine if each graph represents a function. Then find the domain and range. Use interval notation.

▶ **Work with me.**

11.

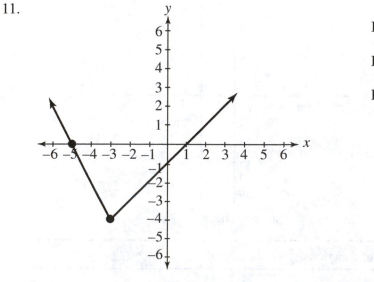

Function?   Yes or No

Domain:

Range:

Work with me.

12.

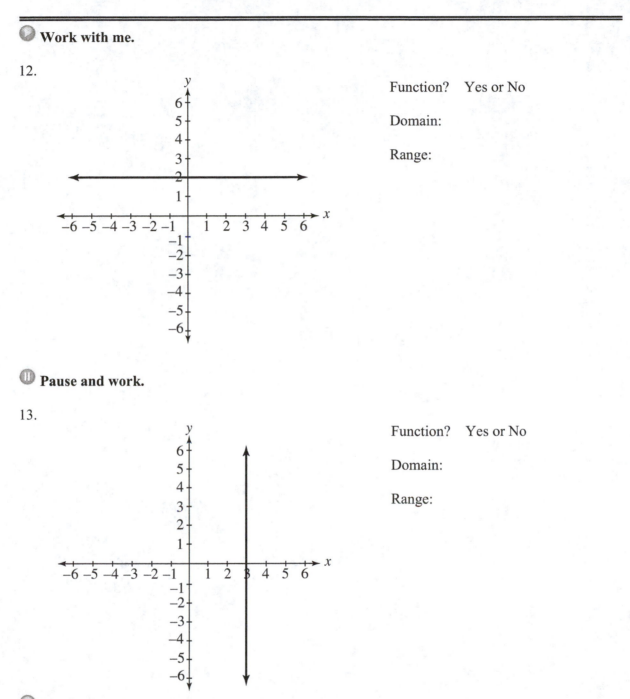

Function?   Yes or No

Domain:

Range:

Pause and work.

13.

Function?   Yes or No

Domain:

Range:

Play and check.

Section 3.6 Functions

Section 4.1 Solving Systems of Linear Equations by Graphing

Complete the outline as you view Video Lecture 4.1. Pause ⏸ the video as needed to fill in the blanks. Then press Play ▶ to continue. Also, circle your answer to each numbered exercise.

Objective 1 **Determine if an ordered pair is a solution of a system of equations in two variables**

A system of linear equations consists of two or more linear equations.

⏸ A _____ of a system consists of an ordered pair that satisfies all equations of the system.

Determine if the given ordered pair is a solution of the following system of equations.

$$\begin{cases} 3x - y = 5 \\ x + 2y = 11 \end{cases}$$

▶ **Work with me.**
1. Is $(3, 4)$ a solution?

⏸ **Pause and work.**
   Is $(0, -5)$ a solution?

▶ Play and check.

Objective 2 **Solve a system of linear equations by graphing**

To solve systems by graphing, look for points in common to the graphs of all equations.

Section 4.1 Solving Systems of Linear Equations by Graphing

Solve the system of equation by graphing.

⬤ **Work with me.**

2. $\begin{cases} 2x + y = 0 \\ 3x + y = 1 \end{cases}$

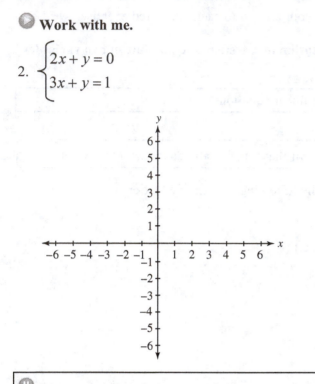

⏸ Intersecting lines have one point in common—the system has _____ _____.

⏸ If equations have different graphs: _____ _____
If a system has at least one solution: _____ _____

Section 4.1 Solving Systems of Linear Equations by Graphing

Solve the system of equations by graphing.

⏸ **Pause and work.**

3. $\begin{cases} x + y = 5 \\ x + y = 6 \end{cases}$

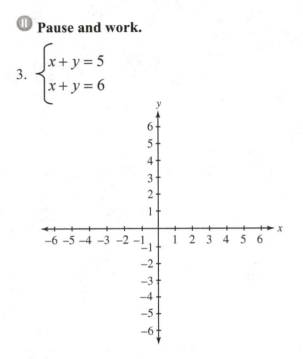

▶ Play and check.

⏸ **Parallel lines:** No point in common—system has _____ _____.

⏸ If two lines have the same slope, but different y-intercepts, the lines are _____.

No solution: inconsistent system
Equations with different graphs: independent equations

Section 4.1 Solving Systems of Linear Equations by Graphing

Solve the system of equations by graphing.

 **Work with me.**

4. $\begin{cases} 6x - y = 4 \\ \dfrac{1}{2}y = -2 + 3x \end{cases}$

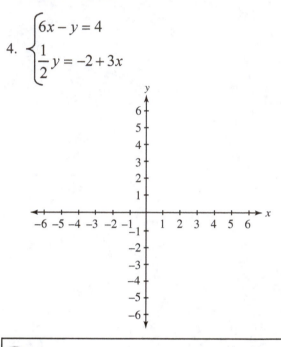

⏸ Same line—infinite number of points in common—the system has an _____ _____ of _____.

⏸ If the lines have the same slope and the same y-intercept, the lines are the _____.

Equations with the same graph: dependent equations
At least one solution: consistent system

⏸ **Review of graphing a system of equations**

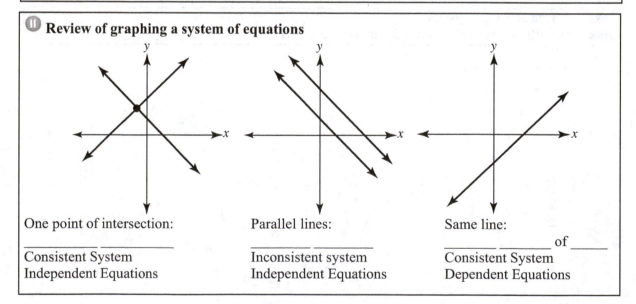

One point of intersection: Parallel lines: Same line:

_____ _____   _____ _____   _____ _____ of _____

Consistent System   Inconsistent system   Consistent System
Independent Equations   Independent Equations   Dependent Equations

Section 4.1 Solving Systems of Linear Equations by Graphing

Objective 3 **Without graphing, determine the number of solutions of a system**

Determine the number of solutions for each system of equations.

Work with me.

5. $\begin{cases} 4x + y = 24 \\ x + 2y = 2 \end{cases}$

Work with me.

6. $\begin{cases} x + y = 4 \\ x + y = 3 \end{cases}$

Pause and work.

7. $\begin{cases} 6y + 4x = 6 \\ 3y - 3 = -2x \end{cases}$

Play and check.

Section 4.1 Solving Systems of Linear Equations by Graphing

Section 4.2 Solving Systems of Linear Equations by Substitution

Complete the outline as you view Video Lecture 4.2. Pause ⏸ the video as needed to fill in the blanks. Then press Play ▶ to continue. Also, circle your answer to each numbered exercise.

(Objective 1) Use the substitution method to solve a system of linear equations

Solve the following system of linear equations in two variables by the substitution method.

▶ **Work with me.**

1. $\begin{cases} x + y = 6 \\ \quad y = -3x \end{cases}$

An ordered pair solution of the system is a solution of each equation of the system.

Solve the following system of linear equations in two variables by the substitution method.

▶ **Work with me.**

2. $\begin{cases} 3x - y = 1 \\ 2x - 3y = 10 \end{cases}$

⏸ **Solving a System of Two Linear Equations by Substitution**
Step 1: Solve one of the equations for _____ of its variables.
Step 2: _____ the expression for the variable found in Step 1 into the _____ equation.
Step 3: _____ the equation from Step 2 to find the value of one variable.
Step 4: _____ the value found in Step 3 in _____ equation containing both variables to find the value of the other variable.
Step 5: _____ the proposed solution in the original system.

Section 4.2 Solving Systems of Linear Equations by Substitution

Solve the following system of linear equations in two variables by the substitution method.

⏵ **Work with me.**

3. $\begin{cases} 3x + 6y = 9 \\ 4x + 8y = 16 \end{cases}$

⏸ Variables subtract out and a false statement: The system has _____ _____.

Solve the following system of linear equations in two variables by the substitution method.

⏵ **Work with me.**

4. $\begin{cases} \dfrac{1}{3}x - y = 2 \\ x - 3y = 6 \end{cases}$

⏸ Variables subtract out and a true statement: The system has an _____ _____ of _____.

Section 4.3 Solving Systems of Linear Equations by Addition

Complete the outline as you view Video Lecture 4.3. Pause ⏸ the video as needed to fill in the blanks. Then press Play ▶ to continue. Also, circle your answer to each numbered exercise.

Objective 1 **Use the addition method to solve a system of linear equations**

Solve the system of equations using the addition method.

▶ **Work with me.**

1. $\begin{cases} x - 2y = 8 \\ -x + 5y = -17 \end{cases}$

A solution of the system is a solution of each equation of the system.

Solve the system of equations using the addition method.

▶ **Work with me.**

2. $\begin{cases} 3x - 2y = 7 \\ 5x + 4y = 8 \end{cases}$

⏸ **Solving a System of Two Linear Equations by the Addition Method**
  Step 1: Rewrite each equation in standard form: _____ + _____ = _____
  Step 2: If necessary, multiply one or both equations by a non-zero number so that the coefficients of a chosen variable in the system are _____.
  Step 3: _____ the equations.
  Step 4: Find the value of one variable by solving the resulting equation from Step 3.
  Step 5: Find the value of the second variable by substituting the value found in Step 4 into either of the original equations.
  Step 6: _____ the proposed solution in the original system.

Section 4.3 Solving Systems of Linear Equations by Addition

Solve the system of equations using the addition method.

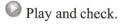 **Work with me.**

3. $\begin{cases} \dfrac{x}{3} - y = 2 \\ \dfrac{-x}{2} + \dfrac{3y}{2} = -3 \end{cases}$

⏸ Variables subtract out and true statement: The system has an _____ _____ of _____.

⏸ Variables subtract out and false statement: The system has _____ _____.

Solve the system using the addition method.

⏸ **Pause and work.**

4. $\begin{cases} 3.5x + 2.5y = 17 \\ -1.5x - 7.5y = -33 \end{cases}$

▶ Play and check.

Section 4.4 Systems of Linear Equations and Problem Solving

Complete the outline as you view Video Lecture 4.4. Pause ⏸ the video as needed to fill in the blanks. Then press Play ▶ to continue. Also, circle your answer to each numbered exercise.

Objective 1 Use a system of equations to solve problems

⏸ **Review the Steps for Problem Solving**
Step 1: _____ the problem.
Step 2: _____ the problem into two equations.
Step 3: _____ the system of equations.
Step 4: _____ the results: Check the proposed solution in the stated problem and state your
conclusion.

▶ **Work with me.**

1. Two numbers total 83 and have a difference of 17. Find the two numbers.

⏸ **Pause and work.**

2. Anne-Marie Jones has been pricing Amtrak train fares for a group trip to New York. Three adults and four children must pay $159. Two adults and three children must pay $112. Find the price of an adult's ticket, and find the price of a child's ticket.

▶ Play and check.

Section 4.4 Systems of Linear Equations and Problem Solving

⏺ **Work with me.**

3. The Santa Fe national historic Trail is approximately 1200 miles between Old Franklin, Missouri, and Santa Fe, New Mexico. Suppose that a group of hikers start from each town and walk the trail toward each other. They meet after a total hiking time of 240 hours. If one group travels $\frac{1}{2}$ mile per hour slower than the other group, find the rate of each group. (Source: National Park Service)

	r ·	t =	d
one group	x		
other group	y		

⏸ After setting up the system, Pause ⏸ the video to solve the system and interpret your answer.

⏺ Play and check.

⏺ **Work with me.**

4. Doreen Schmidt is a chemist with Gemco Pharmaceutical. She needs to prepare 12 liters of a 9% hydrochloric acid solution. Find the amount of a 4% solution and the amount of a 12% solution she should mix to get this solution.

rate	liters of solution	liters of pure acid
0.04	x	
0.12	y	
0.09		

⏸ After setting up the system, Pause ⏸ the video to solve the system and interpret your answer.

⏺ Play and check.

Section 4.4 Systems of Linear Equations and Problem Solving

▶ **Work with me.**

5. Recall that two angles are complementary if the sum of their measures is 90°. Find the measures of two complementary angles if one angle is twice the other.

Section 4.4 Systems of Linear Equations and Problem Solving

Section 4.5 Graphing Linear Inequalities

Complete the outline as you view Video Lecture 4.5. Pause ⏸ the video as needed to fill in the blanks. Then press Play ▶ to continue. Also, circle your answer to each numbered exercise.

Objective 1 **Determine whether an ordered pair is a solution of a linear inequality in two variables**

Linear Equation in two variables: Can be written as $Ax + By = C$ where A, B, and C are real numbers and A and B are not both 0.

Linear Inequality in two variables: Can be written as
$$Ax + By > C$$
$$Ax + By < C$$
$$Ax + By \geq C$$
$$Ax + By \leq C$$
where A, B, and C are real numbers and A and B are not both 0.

▶ **Work with me.**

1. Given the inequality $x < -y$, determine if the following ordered pairs are solutions.
   (0, 2):

⏸ **Pause and work.**

   $(-5, 1)$

▶ Play and check.

⏸ A _____ of an inequality in x and y consists of a value for x and a value for y so that a true statement results.

Section 4.5 Graphing Linear Inequalities

Objective 2 Graph a linear inequality in two variables

Graph the linear inequality in two variables.

 Work with me.

2. $y \geq 2x$

x	y
0	
1	
-2	

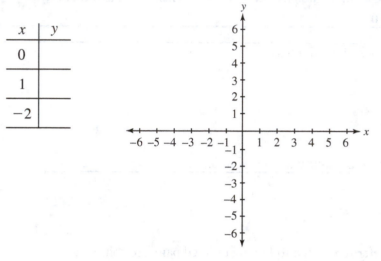

To determine where to shade: Choose a test point <u>not</u> on the boundary line.

Graphing a Linear Inequality in Two Variables

Step 1: Graph the_____ _____ found by replacing the inequality sign with an equal sign. If the inequality sign is $>$ or $<$, then graph a _____ boundary line. If the inequality sign is $\geq$ or $\leq$, graph a _____ boundary line.

Step 2: Choose a point, _____ on the boundary line, as a test point. Substitute the coordinates of this test point into the_____ inequality.

Step 3: If a _____ statement is obtained in Step 2, shade the half-plane that contains the test point. If a _____ statement is obtained, shade the half-plane that does not contain the test point.

Graph linear inequality in two variables.

Work with me.

3. $2x + 7y > 5$

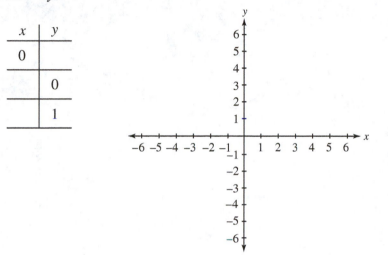

x	y
0	
	0
	1

Complete the outline as you view Video Lecture 4.6. Pause ⏸ the video as needed to fill in the blanks. Then press Play ▶ to continue. Also, circle your answer to each numbered exercise.

⬭Objective 1⬭ Solve a system of linear inequalities

A system of linear inequalities consists of two or more linear inequalities.

⏸ A _____ of a system consists of an ordered pair that satisfies all inequalities of the system.

Graph the solution of the system of linear inequalities.

▶ **Work with me.**

1. $\begin{cases} x \geq 3y \\ x + 3y \leq 6 \end{cases}$

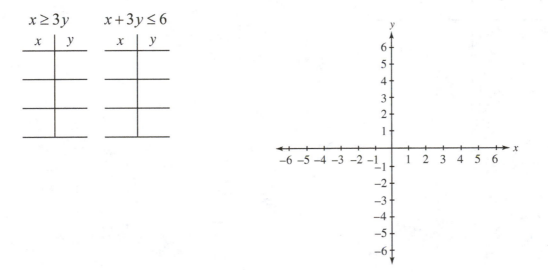

$x \geq 3y$

x	y

$x + 3y \leq 6$

x	y

⏸ **Boundary Lines**
   If $\geq$ or $\leq$, the boundary line is _____.
   If $>$ or $<$, the boundary line is _____.

⏸ Remember, for shading, choose a test point _____ on the boundary line.

⏸ The solution of a system of linear inequalities consists of the _____ of the solution regions.

Section 4.6 Systems of Linear Inequalities

⏸ **Graphing the Solution of a System of Linear Inequalities**
   Step 1: _____ each inequality in the system on the same set of axes.
   Step 2: The solutions of the system are the points common to the graphs of _____ the
              inequalities in the system.

Graph the solution of the system of inequalities.

🔵 **Work with me.**

2. $\begin{cases} y \geq 1 \\ x < -3 \end{cases}$

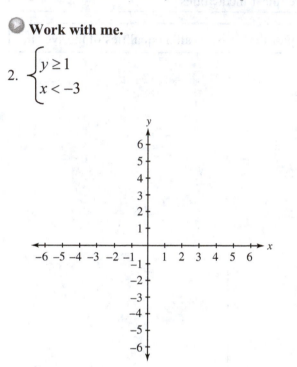

Complete the outline as you view Video Lecture 5.1. Pause ⏸ the video as needed to fill in the blanks. Then press Play ▶ to continue. Also, circle your answer to each numbered exercise.

Objective 1 **Evaluate exponential expressions**

Exponential Notation: a short-hand notation for repeated multiplication of the same factor

---

⏸ **Base:** the repeated _____

  Exponent: number of times the _____ is a _____

Evaluate each exponential expression.

▶ **Work with me.**

1. 2^3

▶ **Work with me.**

2. 3^1

If a number or variable has no exponent, the exponent is an understood 1.

Evaluate each exponential expression.

▶ **Work with me.**

3. $(-4)^2$

▶ **Work with me.**

4. -4^2

Section 5.1 Exponents

⏸ **Pause and work.**

5. $\left(\dfrac{1}{2}\right)^4$

▶ Play and check.

⏸ **Pause and work.**

6. $(0.5)^3$

▶ Play and check.

⏸ **Pause and work.**

7. $4 \cdot 3^2$

▶ Play and check.

Evaluate means "to find the value of."

Evaluate the expression using the given replacement value.

▶ **Work with me.**

8. $\dfrac{2z^4}{5}$; $z = -2$

Objective 2   Use the product rule for exponents

Simplify the expression.

Work with me.

9. $x^2 \cdot x^3$

Product Rule for Exponents

If m and n are positive integers and a is a real number, then

$$a^m \cdot a^n = a^{m+n}$$

Use the product rule to simplify each expression.

Pause and work.

10. $y^3 \cdot y^2 \cdot y^1$

Play and check.

Work with me.

11. $(-5)^7 \cdot (-5)^8$

Work with me.

12. $\left(5y^4\right)(3y)$

Pause and work.

13. $\left(x^9 y\right)\left(x^{10} y^5\right)$

Play and check.

Section 5.1 Exponents

⬭ **Objective 3** Use the power rule for exponents

Use the power rule to simplify the expression.

🔘 **Work with me.**

14. $\left(x^2\right)^3$

Power Rule for Exponents

If m and n are positive integers and a is a real number, then

$$\left(a^m\right)^n = a^{mn}$$

Use the power rule to simplify each expression.

⏸ **Pause and work.**

15. $\left(x^9\right)^4$

▶ Play and check.

🔘 **Work with me.**

16. $\left[\left(-5\right)^3\right]^7$

⬭ **Objective 4** Use the power rules for products and quotients

---

Power of a Product Rule

If n is a positive integer and a and b are real numbers, then

$$\left(ab\right)^n = a^n b^n$$

Use the power of a product rule to simplify the expression.

▶ **Work with me.**

17. $\left(pq \right)^{8}$

Power of a Quotient Rule

If n is a positive integer and a and c are real numbers, then

$$\left(\frac{a}{c} \right)^{n} = \frac{a^{n}}{c^{n}}, \quad c \neq 0$$

Use the power of a quotient rule to simplify each expression.

▶ **Work with me.**

18. $\left(\dfrac{r}{s} \right)^{9}$

⏸ **Pause and work.**

19. $\left(x^{2}y^{3} \right)^{5}$

▶ Play and check.

▶ **Work with me.**

20. $\left(\dfrac{-2xz}{y^{5}} \right)^{2}$

$\boxed{\text{Objective 5}}$ **Use the quotient rule for exponents and define a number raised to the 0 power**

Simplify the expression.

🔘 **Work with me.**

21. $\dfrac{x^5}{x^2}$

Quotient Rule for Exponents
If m and n are positive integers and a is a real number, then
$$\dfrac{a^m}{a^n} = a^{m-n}$$
as long as a is not 0.

Simplify each expression.

⏸ **Pause and work.**

22. $\dfrac{(-4)^6}{(-4)^3}$

🔘 Play and check.

🔘 **Work with me.**

23. $\dfrac{9a^4b^7}{27ab^2}$

▶ **Work with me.**

24. $\dfrac{x^3}{x^3}$

Zero Exponent

$$a^0 = 1$$

as long as a is not 0.

Simplify each expression.

▶ **Work with me.**

25. $(2x)^0$

▶ **Work with me.**

26. $-7x^0$

⏸ **Pause and work.**

27. $5^0 + y^0$

▶ Play and check.

⬭ **Objective 6** Decide which rule(s) to use to simplify an expression

Simplify each expression.

▶ **Work with me.**

28. $(2x^3)(-8x^4)$

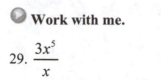 **Work with me.**

29. $\dfrac{3x^5}{x}$

Pause and work.

30. $\left(\dfrac{3y^5}{6x^4}\right)^3$

▶ Play and check.

Complete the outline as you view Video Lecture 5.2. Pause ⏸ the video as needed to fill in the blanks. Then press Play ▶ to continue. Also, circle your answer to each numbered exercise.

⬭ Objective 1 Define polynomial, monomial, binomial, trinomial and degree

⏸ Remember, a _____ is a number or the product of numbers and variables raised to powers.

A <u>polynomial</u> in x is a finite sum of terms of the form

$$\text{real number} \longrightarrow ax^{n} \longleftarrow \text{whole number}$$

⏸ A _____ is a polynomial with exactly 1 term.

A _____ is a polynomial with exactly 2 terms.

A _____ is a polynomial with exactly 3 terms.

Degree of a Term

The degree of a term is the <u>sum of the exponents</u> on the variables contained in the term.

Degree of a Polynomial

The degree of a polynomial is the <u>greatest</u> degree <u>of any term</u> of the polynomial.

Determine if each polynomial is a monomial, binomial, trinomial, or polynomial with four or more terms. Then find the degree of the polynomial.

▶ Work with me.

1. $-2t^2 + 3t + 6$

   type of polynomial:

   terms:

   degree of each term:

   degree of polynomial:

Section 5.2 Adding and Subtracting Polynomials

⏸ **Pause and work.**

2. $12x^4y - x^2y^2 - 12x^2y^4$

▶ Play and check.

(Objective 2) **Find the value of a polynomial given replacement values for the variables**

▶ **Work with me.**

3. Find the value of $x^2 - 5x - 2$ when $x = 0$ and $x = -1$.

(Objective 3) **Simplify a polynomial by combining like terms**

⏸ Like terms contain the _____ variables raised to exactly the _____ powers.

Simplify by combining like terms.

▶ **Work with me.**

4. $14x^2 + 9x^2$

▶ **Work with me.**

5. $15x^2 - 3x^2 - y$

⏸ **Pause and work.**

6. $0.1y^2 - 1.2y^2 + 6.7 - 1.9$

▶ Play and check.

Objective 4   **Add and subtract polynomials**

> **Adding Polynomials**
> To add polynomials, combine all like terms.

Add the polynomials.

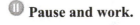 **Work with me.**

7. $(-7x+5)+\left(-3x^2+7x+5\right)$

Use a vertical format to add the polynomials.

⏸ **Pause and work.**

8. $\begin{aligned} 3t^2+4 \\ +5t^2-8 \end{aligned}$

▶ Play and check.

> **Subtracting Polynomials**
> To subtract two polynomials, change the signs of the terms of the polynomial being subtracted and then add.

Subtract the polynomials.

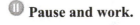 **Work with me.**

9. $\left(2x^2+5\right)-\left(3x^2-9\right)$

Section 5.2 Adding and Subtracting Polynomials

Use a vertical format to subtract the polynomials.

⏵ **Work with me.**

10.
$$5x^3 - 4x^2 + 6x - 2$$
$$-\left(3x^3 - 2x^2 - x - 4\right)$$

⏸ **Pause and work.**

11. Subtract $\left(19x^2 + 5\right)$ from $\left(81x^2 + 10\right)$.

⏵ Play and check.

Complete the outline as you view Video Lecture 5.3. Pause ⏸ the video as needed to fill in the blanks. Then press Play ▶ to continue. Also, circle your answer to each numbered exercise.

Objective 1 **Multiply monomials**

> **Commutative Property:** may reorder factors
>
> **Associative Property:** may regroup factors

> **Product Rule for Exponents:** $a^m \cdot a^n = a^{m+n}$

Multiply.

▶ **Work with me.**

1. $\left(-\dfrac{1}{3}y^2\right)\left(\dfrac{2}{5}y\right)$

Objective 2 **Use the distributive property to multiply polynomials**

Multiply.

▶ **Work with me.**

2. $3x(2x+5)$

> **Distributive Property:** $a(b+c) = a \cdot b + a \cdot c$

Section 5.3 Multiplying Polynomials

Multiply.

⏸ **Pause and work.**

3. $-y\left(4x^3 - 7x^2 y + xy^2 + 3y^3\right)$

▶ Play and check.

To Multiply Two Polynomials

Multiply each term of the first polynomial by each term of the second polynomial, and then combine terms.

Multiply.

▶ **Work with me.**

4. $(a+7)(a-2)$

▶ **Work with me.**

5. $\left(3x^2 + 1\right)^2$

⏸ **Pause and work.**

6. $(x+5)\left(x^3 - 3x + 4\right)$

▶ Play and check.

Objective 3   **Multiply polynomials vertically**

Multiply.

Work with me.

7. $(5x+1)(2x^2+4x-1)$

Rewrite: $\begin{array}{r} 2x^2+4x-1 \\ 5x+1 \\ \hline \end{array}$

Section 5.3 Multiplying Polynomials

Section 5.4 Special Products

Complete the outline as you view Video Lecture 5.4. Pause ⏸ the video as needed to fill in the blanks. Then press Play ▶ to continue. Also, circle your answer to each numbered exercise.

Objective 1 **Multiply two binomials using the FOIL method**

FOIL Order used: binomial · binomial

FOIL Order: First Outside Inside Last

Multiply using the FOIL method.

▶ **Work with me.**

1. $(x+3)(x+4)$

⏸ **Pause and work.**

2. $(3b+7)(2b-5)$

▶ Play and check.

Multiply.

▶ **Work with me.**

3. $\left(5x+9\right)^2$

Section 5.4 Special Products

Objective 2 Square a binomial

Squaring a Binomial

A binomial squared is equal to the square of the first term plus or minus twice the product of both terms plus the square of the second term.

$$(a+b)^2 = a^2 + 2ab + b^2$$
$$(a-b)^2 = a^2 - 2ab + b^2$$

Multiply.

❚❚ Pause and work.

4. $(2x-1)^2$

▶ Play and check.

Objective 3 Multiply the sum and difference of two terms

Multiply.

▶ **Work with me.**

5. $(9x+y)(9x-y)$

Sum and Difference of Same Two Terms

$$(a+b)\ (a-b)$$

Multiplying the Sum and Difference of Two Terms

The product of the sum and difference of two terms is the square of the first term minus the square of the second term.

$$(a+b)\ (a-b) = a^2 - b^2$$

124

Multiply.

⏵ **Work with me.**

6. $(a-7)(a+7)$

⏸ **Pause and work.**

7. $(4x+5)(4x-5)$

⏵ Play and check.

⏵ **Work with me.**

8. $\left(\dfrac{1}{3}a^2 - 7\right)\left(\dfrac{1}{3}a^2 + 7\right)$

(**Objective 4**) **Use special products to multiply**

Multiply.

⏵ **Work with me.**

9. $(3a+1)^2$

⏵ **Work with me.**

10. $(x+3)\left(x^2 - 6x + 1\right)$

Section 5.4 Special Products

Complete the outline as you view Video Lecture 5.5. Pause ⏸ the video as needed to fill in the blanks. Then press Play ▶ to continue. Also, circle your answer to each numbered exercise.

Objective 1   **Simplify expressions containing negative exponents**

Quotient Rule: $\dfrac{a^m}{a^n} = a^{m-n}$

Negative Exponents

If a is a real number other than 0 and n is an integer, then $a^{-n} = \dfrac{1}{a^n}$.

A negative exponent has nothing to do with the sign of the base.

Evaluate.

▶ **Work with me.**

1. 3^{-2}

Simplify the expression. Write the result with positive exponents only.

▶ **Work with me.**

2. $2x^{-3}$

Evaluate.

⏸ **Pause and work.**

3. $(-2)^{-4}$

▶ Play and check.

Section 5.5 Negative Exponents and Scientific Notation

Simplify the expression. Write the result with positive exponents only.

⏵ **Work with me.**

4. $\dfrac{1}{y^{-4}}$

Negative Exponents: $\dfrac{1}{a^{-n}} = a^n$

Evaluate.

⏵ **Work with me.**

5. $2^{-1} + 4^{-1}$

⏸ **Pause and work.**

6. $\dfrac{1}{7^{-2}}$

⏵ Play and check.

Negative Exponents

If a is a real number other than 0 and n is an integer, then $a^{-n} = \dfrac{1}{a^n}$ or $\dfrac{1}{a^{-n}} = a^n$.

(**Objective 2**)   **Use all the rules and definitions for exponents to simplify exponential expressions**

Simplify the expression. Write the result with positive exponents only.

⏵ **Work with me.**

7. $\dfrac{r}{r^{-3}r^{-2}}$

Section 5.5 Negative Exponents and Scientific Notation

Work with me.

8. $\dfrac{\left(-2xy^{-3}\right)^{-3}}{\left(xy^{-1}\right)^{-1}}$

Scientific Notation

A positive number is written in scientific notation if it is written as the product of a number a, where $1 \le a < 10$, and an integer power r of 10: $a \times 10^r$

Objective 3   **Write numbers in scientific notation**

To Write a Number in Scientific Notation:

Step 1: Move the decimal point in the original number to the left or right so that the new number has a value between _____ and _____.

Step 2: Count the number of decimal places the decimal point is moved in Step 1. If the original number is 10 or _____, the count is _____. If the original number is less than 1, the count is _____.

Step 3: Multiply the new number in Step 1 by 10 raised to an exponent equal to the count found in Step 2.

Write each number in scientific notation.

Work with me.

9. 78,000

Pause and work.

10. 0.00000167

Play and check.

Section 5.5 Negative Exponents and Scientific Notation

Objective 4 **Convert numbers from scientific notation to standard form**

Write each number in standard form.

▶ **Work with me.**

11. 3.3×10^{-2}

⏸ **Pause and work.**

12. 2.032×10^{4}

▶ Play and check.

Objective 5 **Perform operations on numbers written in scientific notation**

Evaluate the expression. Write the result in standard form.

▶ **Work with me.**

13. $\dfrac{1.4 \times 10^{-2}}{7 \times 10^{-8}}$

Section 5.6 Dividing Polynomials

Complete the outline as you view Video Lecture 5.6. Pause ⏸ the video as needed to fill in the blanks. Then press Play ▶ to continue. Also, circle your answer to each numbered exercise.

Objective 1 **Divide a polynomial by a monomial**

> **Dividing a Polynomial by a Monomial**
>
> Divide each term of the polynomial by the monomial.
>
> $$\frac{a+b}{c} = \frac{a}{c} + \frac{b}{c}, \ c \neq 0$$

Divide.

▶ **Work with me.**

1. $\dfrac{12x^4 + 3x^2}{x}$

⏸ **Pause and work.**

2. $\dfrac{-9x^5 + 3x^4 - 12}{3x^3}$

▶ Play and check.

Objective 2 **Use long division to divide a polynomial by another polynomial**

> Dividing by a polynomial with two or more terms: long division

> **Fractional Part of Quotient:** $\dfrac{\text{Remainder}}{\text{Divisor}}$

> **Long division steps:** Divide, Multiply, Subtract, Bring Down

Section 5.6 Dividing Polynomials

Divide using long division.

🔘 **Work with me.**

3. $\dfrac{x^2 + 4x + 3}{x + 3}$

⏸ **To subtract:** _____ the signs of ALL terms being subtracted.

Divide.

🔘 **Work with me.**

4. $\dfrac{2b^3 + 9b^2 + 6b - 4}{b + 4}$

🔘 **Work with me.**

5. $\dfrac{x^3 - 27}{x - 3}$

If missing powers in dividend or divisor, fill in with a coefficient of zero.

Section 6.1 The Greatest Common Factor and Factoring by Grouping

Complete the outline as you view Video Lecture 6.1. Pause ⏸ the video as needed to fill in the blanks. Then press Play ▶ to continue. Also, circle your answer to each numbered exercise.

Objective 1 **Find the greatest common factor of a list of integers**

> ⏸ **Finding the GCF (Greatest Common Factor) of a List of Integers**
>
> **Step 1:** Write each number as a product of _____ numbers.
>
> **Step 2:** Identify the _____ prime factors.
>
> **Step 3:** The product of all common prime factors found in Step 2 is the _____
>
> _____ _____. If there are no common prime factors, the greatest
>
> common factor is 1.

▶ **Work with me.**

1. Find the GCF of 36 and 90.

> ⏸ A _____ _____ is a natural number other than 1 whose only factors are 1 and
>
> itself.

Objective 2 **Find the greatest common factor of a list of terms**

▶ **Work with me.**

2. Find the GCF of x^2, x^3 and x^5.

> The GCF of common variable factors is the variable raised to the smallest exponent.

⏸ **Pause and work.**

3. Find the GCF of $12y^4$ and $20y^3$.

▶ Play and check.

Section 6.1 The Greatest Common Factor and Factoring by Grouping

Objective 3   **Factor out the greatest common factor from a polynomial**

Factor out the GCF.

▶ **Work with me.**

4. $30x - 15$

⏸ **Pause and work.**

5. $14x^3y + 7x^2y - 7xy$

▶ Play and check.

⏸ Always check factoring by _____.

⏸ FACTORING means _____ _____ _____ _____.

Factor out the GCF.

▶ **Work with me.**

6. $y(x^2 + 2) + 3(x^2 + 2)$

Objective 4   **Factor a polynomial by grouping**

Factor the polynomial.

▶ **Work with me.**

7. $5xy - 15x - 6y + 18$

Section 6.1 The Greatest Common Factor and Factoring by Grouping

---

Ⅱ To Factor by Grouping

Step 1: Group the terms in two groups of two terms so that each group has a _____

    factor.

Step 2: Factor out the _____ from each group.

Step 3: If there is now a common binomial factor in the groups, _____ it out.

Step 4: If not, _____ the terms and try these steps again.

Factor the polynomial by grouping.

Ⅱ Pause and work.

8. $6a^2 + 9ab^2 + 6ab + 9b^3$

▶ Play and check.

Section 6.1 The Greatest Common Factor and Factoring by Grouping

Section 6.2 Factoring Trinomials of the Form $x^2 + bx + c$

Complete the outline as you view Video Lecture 6.2. Pause ⏸ the video as needed to fill in the blanks. Then press Play ▶ to continue. Also, circle your answer to each numbered exercise.

Objective 1 **Factor trinomials of the form $x^2 + bx + c$**

⏸ Factoring means _____ _____ _____ _____ .

Factoring a Trinomial of the Form $x^2 + bx + c$

The product of $x^2 + bx + c$ is

The product of these numbers is c.

$$x^2 + bx + c = (x + \boxed{}) \, (x + \boxed{})$$

The sum of these numbers is b.

Factor the trinomial.

▶ **Work with me.**

1. $x^2 + 7x + 6$ Find two numbers whose product is 6 and whose sum is 7.

The order of the factors makes no difference because multiplication is commutative.

Factor each trinomial.

▶ **Work with me.**

2. $x^2 - 8x + 15$ Find two numbers whose product is 15 and sum is -8.

Section 6.2 Factoring Trinomials of the Form $x^2 + bx + c$

⏸ **Pause and work.**

3. $x^2 - 3x - 18$

▶ Play and check.

▶ **Work with me.**

4. $x^2 - 3xy - 4y^2$

(Objective 2) **Factor out the greatest common factor and then factor a trinomial of the form**

$x^2 + bx + c$

Factor each trinomial.

▶ **Work with me.**

5. $3x^2 + 9x - 30$

⏸ **Pause and work.**

6. $5x^3y - 25x^2y^2 - 120xy^3$

▶ Play and check.

Section 6.3 Factoring Trinomials of the Form $ax^2 + bx + c$ and Perfect Square Trinomials

Complete the outline as you view Video Lecture 6.3. Pause ⏸ the video as needed to fill in the blanks. Then press Play ▶ to continue. Also, circle your answer to each numbered exercise.

Objective 1 **Factor trinomials of the form $ax^2 + bx + c$, where $a \neq 1$**

Factor.

▶ **Work with me.**

1. $10x^2 + 31x + 3$

⏸ **Pause and work.**

2. $4x^2 - 8x - 21$

▶ Play and check.

Middle Term = Outside Product + Inside Product

Objective 2 **Factor out a GCF before factoring a trinomial of the form $ax^2 + bx + c$**

Factor each trinomial.

▶ **Work with me.**

3. $30x^3 + 38x^2 + 12x$

Section 6.3 Factoring Trinomials of the Form $ax^2 + bx + c$ and Perfect Square Trinomials

⏸ Pause and work.

4. $4x^3 - 9x^2 - 9x$

▶ Play and check.

▶ Work with me.

5. $-14x^2 + 39x - 10$

▶ Work with me.

6. $x^2 + 14x + 49$

Perfect square trinomial — factors as a binomial squared.

Factoring Perfect Square Trinomials
$$a^2 + 2ab + b^2 = (a + b)^2$$ $$a^2 - 2ab + b^2 = (a - b)^2$$

Section 6.3 Factoring Trinomials of the Form $ax^2 + bx + c$ and Perfect Square Trinomials

Objective 3 **Factor perfect square trinomials**

Factor each perfect square trinomial.

Work with me.

7. $x^2 + 22x + 121$

Pause and work.

8. $9x^2 - 24xy + 16y^2$

Play and check.

Section 6.3 Factoring Trinomials of the Form $ax^2 + bx + c$ and Perfect Square Trinomials

Section 6.4 Factoring Trinomials of the Form $ax^2 + bx + c$ by Grouping

Complete the outline as you view Video Lecture 6.4. Pause ⏸ the video as needed to fill in the blanks. Then press Play ▶ to continue. Also, circle your answer to each numbered exercise.

Objective 1   Use the grouping method to factor trinomials of the form $ax^2 + bx + c$

Factor by grouping.

▶ **Work with me.**

1. $x^2 + 3x + 2x + 6$

⏸ **To Factor Trinomials by Grouping**

Step 1: Factor out a _____ _____ _____ if there is one other than 1.

Step 2: For the resulting trinomial $ax^2 + bx + c$, find two numbers whose product is

_____ and whose sum is _____.

Step 3: Write the _____ term, bx, using the factors found in Step 2.

Step 4: Factor by _____.

Factor by grouping.

▶ **Work with me.**

2. $21y^2 + 17y + 2$

Section 6.4 Factoring Trinomials of the Form $ax^2 + bx + c$ by Grouping

⏸ **Pause and work.**

3. $10x^2 - 9x + 2$

▶ Play and check.

▶ **Work with me.**

4. $12x^3 - 27x^2 - 27x$

Complete the outline as you view Video Lecture 6.5. Pause ⏸ the video as needed to fill in the blanks. Then press Play ▶ to continue. Also, circle your answer to each numbered exercise.

Objective 1 **Factor the difference of squares**

⏸ A _____ contains two terms.

Factoring the Difference of Two Squares

$$a^2 - b^2 = (a + b)(a - b)$$

Factor each binomial.

▶ **Work with me.**

1. $x^2 - 4$

⏸ **Pause and work.**

2. $121m^2 - 100n^2$

▶ Play and check.

▶ **Work with me.**

3. $16r^2 + 1$

⏸ The <u>sum</u> of two squares <u>cannot</u> be factored.

The sum of two squares is a _____ polynomial.

Section 6.5 Factoring Binomials

Factor each binomial.

⏸ **Pause and work.**

4. $xy^3 - 9xyz^2$

▶ Play and check.

▶ **Work with me.**

5. $49 - \dfrac{9}{25}m^2$

(**Objective 2**) **Factor the sum or difference of two cubes**

Factoring the sum or Difference of Two Cubes
$$a^3 + b^3 = (a+b)(a^2 - ab + b^2)$$ $$a^3 - b^3 = (a-b)(a^2 + ab + b^2)$$

Factor each sum or difference of two cubes.

▶ **Work with me.**

6. $x^3 + 125$

▶ **Work with me.**

7. $x^3 y^3 - 64$

⏸ **Pause and work.**

8. $8m^3 + 64$

▶ Play and check.

Section 6.5 Factoring Binomials

Complete the outline as you view the video lecture for Chapter 6 Integrated Review. Pause ⏸ the video as needed to fill in the blanks. Then Press play ▶ to continue listening.

Objective 1 **Choosing a factoring strategy**

⏸ **Factoring a Polynomial**

Step 1: Are there any common factors? If so, factor out the _____.

Step 2: How many terms are in the polynomial?

    a. If there are two terms, decide if one of the following can be applied.

        i. Difference of _____ squares:

$$a^2 - b^2 = \text{_____}$$

        ii. Difference of two cubes:

$$a^3 - b^3 = \text{_____}$$

        iii. Sum of two cubes:

$$a^3 + b^3 = \text{_____}$$

    b. If there are _____ terms, try one of the following.

        i. Perfect square trinomial:

$$a^2 + 2ab + b^2 = \text{_____}$$
$$a^2 - 2ab + b^2 = \text{_____}$$

        ii. If not a perfect square trinomial, factor using the methods presented in Sections 6.2 – 6.4.

    c. If there are _____ or more terms, try factoring by _____.

Step 3: See if any factors in the factored polynomial can be factored further.

Step 4: Check by _____.

Integrated Review – Choosing a Factoring Strategy

Factor each polynomial completely.

❙❙ Pause and work.

1. $4x^2 - 8xy - 3x + 6y$

▶ Play and check.

❙❙ Pause and work.

2. $125 - 8y^3$

▶ Play and check.

❙❙ Pause and work.

3. $4x^3 + 20x^2 - 56x$

▶ Play and check.

❙❙ Pause and work.

4. $2xy - 72x^3y$

▶ Play and check.

Section 6.6 Solving Quadratic Equations by Factoring

Complete the outline as you view Video Lecture 6.6. Pause ⏸ the video as needed to fill in the blanks. Then press Play ▶ to continue. Also, circle your answer to each numbered exercise.

Objective 1 Solve quadratic equations by factoring

Quadratic Equation

A quadratic equation is one that can be written in the form

$$ax^2 + bx + c = 0$$

where a, b and c are real numbers and $a \neq 0$.

Zero Factor Theorem

If a and b are real numbers and if $ab = 0$, then $a = 0$ or $b = 0$.

Solve the equation.

▶ **Work with me.**

1. $x^2 + 2x - 8 = 0$

⏸ To use the Zero Factor Theorem, one side of the equation must be a _____

_____ and the other side must be _____.

Section 6.6 Solving Quadratic Equations by Factoring

Solve the equation.

⏸ **Pause and work.**

2. $(2x + 3)(4x - 5) = 0$

▶ Play and check.

⏸ A _____ is a number that makes the equation a true statement.

⏸ **To Solve a Quadratic Equation by Factoring**

Step 1: Write the equation in _____ _____ so that one side of the equation is 0.

Step 2: _____ the quadratic equation completely.

Step 3: Set each factor containing a _____ equal to zero.

Step 4: _____ the resulting equations.

Step 5: _____ each solution in the original equation.

Solve the equation.

▶ **Work with me.**

3. $x(3x - 1) = 14$

Objective 2 Solve equations with degree greater than 2 by factoring

Solve the equation.

▶ **Work with me.**

4. $(2x+3)(2x^2-5x-3)=0$

Objective 3 Find the *x*-intercepts of the graph of a quadratic equation in two variables

Find the *x*-intercepts of the graph of the equation.

▶ **Work with me.**

5. $y=2x^2+11x-6$

Section 6.6 Solving Quadratic Equations by Factoring

Section 6.7 Quadratic Equations and Problem Solving

Complete the outline as you view Video Lecture 6.7. Pause ⏸ the video as needed to fill in the blanks. Then press Play ▶ to continue. Also, circle your answer to each numbered exercise.

(**Objective 1**) **Solve problems that can be modeled by quadratic equations**

⏸ Remember the steps for solving word problems.

Step 1: _____

Step 2: _____

Step 3: _____

Step 4: _____

▶ **Work with me.**

1. The perimeter of a triangle is 85 feet. Find the length of its sides.

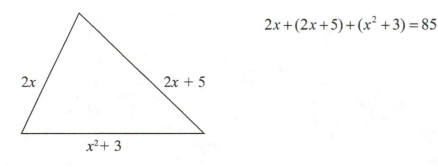

$$2x + (2x + 5) + (x^2 + 3) = 85$$

▶ **Work with me.**

2. An object is thrown upward from the top of an eighty-foot building, with an initial velocity of sixty-four feet per second. The height, h, of the object after t seconds is given by the quadratic equation: $h = -16t^2 + 64t + 80$. When will the object hit the ground?

Section 6.7 Quadratic Equations and Problem Solving

Pythagorean Theorem

In a right triangle, the sum of the squares of the lengths of the two legs is equal to the square of the length of the hypotenuse.

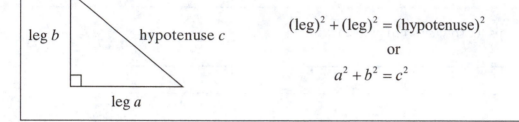

$$(\text{leg})^2 + (\text{leg})^2 = (\text{hypotenuse})^2$$

or

$$a^2 + b^2 = c^2$$

▶ **Work with me.**

3. One leg of a right triangle is 4 millimeters longer than the smaller leg and the hypotenuse is 8 millimeters longer than the smaller leg. Find the lengths of the sides of the triangle.

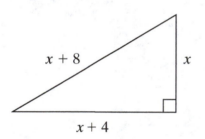

⏸ Pause and work.

4. The sum of a number and its square is 182. Find the number(s).

▶ Play and check.

▶ **Work with me.**

5. The product of two consecutive page numbers is 420. Find the page numbers.

Section 6.7 Quadratic Equations and Problem Solving

Complete the outline as you view Video Lecture 7.1. Pause ⏸ the video as needed to fill in the blanks. Then press Play ▶ to continue. Also, circle your answer to each numbered exercise.

Objective 1) **Find the value of a rational expression given a replacement number**

> **Rational Expression:** $\dfrac{P(\text{polynomial})}{Q(\text{polynomial})}$; $Q \neq 0$

> Rational expressions have different values depending on replacement values.

> ⏸ Evaluate means "_____ _____ _____ _____ _____".

▶ **Work with me.**

1. Evaluate the rational expression when $z = -5$.

$$\dfrac{z}{z^2 - 5}$$

> A rational expression is an expression that can be written in the form $\dfrac{P}{Q}$, where
>
> P and Q are polynomials and $Q \neq 0$.

Objective 2) **Identify values for which a rational expression is undefined**

Find any numbers for which each expression is undefined.

▶ **Work with me.**

2. $\dfrac{x+3}{x+2}$

Section 7.1 Simplifying Rational Expressions

⏵ **Work with me.**

3. $\dfrac{11x^2 + 1}{x^2 - 5x - 14}$

⬭**Objective 3** ⬭ **Simplify or write rational expressions in lowest terms**

⏸ **To Simplify a Rational Expression**

   Step 1: Completely _____ the numerator and denominator.

   Step 2: Divide out _____ common to the numerator and denominator. (This is the same as "removing a factor of 1.")

Simplify each rational expression.

⏵ **Work with me.**

4. $\dfrac{-5a - 5b}{a + b}$

⏸ **Pause and work.**

5. $\dfrac{x + 7}{7 + x}$

⏵ Play and check.

⏵ **Work with me.**

6. $\dfrac{x - 7}{7 - x}$

Section 7.1 Simplifying Rational Expressions

Addition is commutative, subtraction is not.

$$\frac{a-b}{b-a} = -1 \text{ (denominator not 0)}$$

Simplify each rational expression.

Ⅱ Pause and work.

7. $\dfrac{x^3 + 7x^2}{x^2 + 5x - 14}$

▶ Play and check.

▶ Work with me.

8. $\dfrac{4 - x^2}{x - 2}$

Objective 4 Write equivalent rational expressions of the form $-\dfrac{a}{b} = \dfrac{-a}{b} = \dfrac{a}{-b}$

Write 2 equivalent forms of the given rational expression.

▶ Work with me.

9. $-\dfrac{x + 11}{x - 4}$

Section 7.1 Simplifying Rational Expressions

Section 7.2 Multiplying and Dividing Rational Expressions

Complete the outline as you view Video Lecture 7.2. Pause ⏸ the video as needed to fill in the blanks. Then press Play ▶ to continue. Also, circle your answer to each numbered exercise.

Objective 1 **Multiply rational expressions**

Multiply.

▶ **Work with me.**

1. $\dfrac{8x}{2} \cdot \dfrac{x^5}{4x^2}$

Multiplying Rational Expressions

If $\dfrac{P}{Q}$ and $\dfrac{R}{S}$ are rational expressions, then $\dfrac{P}{Q} \cdot \dfrac{R}{S} = \dfrac{PR}{QS}$.

⏸ **Multiplying Rational Expressions**

  Step 1: Completely _____ numerators and denominators.

  Step 2: Multiply _____ and multiply _____.

  Step 3: _____ or write the product in lowest terms by dividing out common factors.

Multiply.

▶ **Work with me.**

2. $\dfrac{5x-20}{3x^2+x} \cdot \dfrac{3x^2+13x+4}{x^2-16}$

Section 7.2 Multiplying and Dividing Rational Expressions

⬭ **Objective 2** **Divide rational expressions**

Dividing Rational Expressions

If $\dfrac{P}{Q}$ and $\dfrac{R}{S}$ are rational expressions, and $\dfrac{R}{S}$ is not 0, then

$$\frac{P}{Q} \div \frac{R}{S} = \frac{P}{Q} \cdot \frac{S}{R} = \frac{PS}{QR}$$

⏸ To divide by a rational expression — _____ by its _____ .

Divide.

▶ **Work with me.**

3. $\dfrac{x+2}{7-x} \div \dfrac{x^2-5x+6}{x^2-9x+14}$

⬭ **Objective 3** **Multiply or divide rational expressions**

Divide.

⏸ **Pause and work.**

4. $\dfrac{5x-10}{12} \div \dfrac{4x-8}{8}$

▶ Play and check.

Find the area of the rectangle.

🔘 **Work with me.**

5.

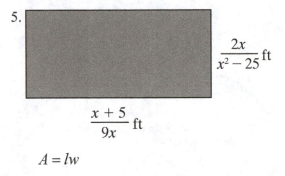

$\dfrac{2x}{x^2 - 25}$ ft

$\dfrac{x + 5}{9x}$ ft

$A = lw$

⬭ **Objective 4** ⬭ **Convert between units of measure**

Convert the units of measure.

🔘 **Work with me.**

6. 3 cubic yards = _____ cubic feet

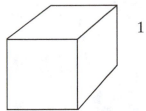

1 cubic yard

Section 7.2 Multiplying and Dividing Rational Expressions

Section 7.3 Adding and Subtracting Rational Expressions with
Common Denominators and Least Common Denominator

Complete the outline as you view Video Lecture 7.3. Pause ⏸ the video as needed to fill in the blanks. Then press Play ▶ to continue. Also, circle your answer to each numbered exercise.

Objective 1 Add and subtract rational expressions with the same denominator

Add the rational expressions.

▶ **Work with me.**

1. $\dfrac{1}{9} + \dfrac{4}{9} =$ $\dfrac{1}{x+7} + \dfrac{4}{x+7} =$

Adding and Subtracting Rational Expressions with Common Denominators

If $\dfrac{P}{R}$ and $\dfrac{Q}{R}$ are rational expressions, then $\dfrac{P}{R} + \dfrac{Q}{R} = \dfrac{P+Q}{R}$ and $\dfrac{P}{R} - \dfrac{Q}{R} = \dfrac{P-Q}{R}$.

Add.

⏸ **Pause and work.**

2. $\dfrac{9}{3+y} + \dfrac{y+1}{3+y}$

▶ Play and check.

Can divide out common _____.

Cannot divide out common _____.

Section 7.3 Adding and Subtracting Rational Expressions with
Common Denominators and Least Common Denominator

To simplify:

Step 1: Factor _____ and _____ .

Step 2: _____ _____ common factors.

Subtract.

Work with me.

3. $\dfrac{2x+3}{x^2-x-30} - \dfrac{x-2}{x^2-x-30}$

Objective 2   Find the least common denominator of a list of rational expressions

---

Finding the Least Common Denominator (LCD)

Step 1: Factor each _____ completely.

Step 2: The least common denominator (LCD) is the _____ of all unique factors found in
Step 1, each raised to a power equal to the greatest number of times that the factor
appears in any one factored denominator.

Find the least common denominator (LCD).

Work with me.

4. $\dfrac{9}{8x}, \dfrac{3}{2x+4}$

$8x =$

$2x+4 =$

$LCD =$

Section 7.3 Adding and Subtracting Rational Expressions with
Common Denominators and Least Common Denominator

❚❚ Pause and work.

5. $\dfrac{1}{3x+3}, \dfrac{8}{2x^2+4x+2}$

$$3x+3 =$$

$$2x^2+4x+2 =$$

$$LCD =$$

▶ Play and check.

(**Objective 3**) **Write a rational expression as an equivalent expression whose denominator is given**

Write an equivalent rational expression with the given denominator.

▶ **Work with me.**

6. $\dfrac{6}{3a} = \dfrac{}{12ab^2}$

▶ **Work with me.**

7. $\dfrac{9a+2}{5a+10} = \dfrac{}{5b(a+2)}$

| Equivalent rational expressions simplify to the same rational expression. |

Section 7.3 Adding and Subtracting Rational Expressions with
Common Denominators and Least Common Denominator

Section 7.4 Adding and Subtracting Rational Expressions with Unlike Denominators

Complete the outline as you view Video Lecture 7.4. Pause ⏸ the video as needed to fill in the blanks. Then press Play ▶ to continue. Also, circle your answer to each numbered exercise.

(**Objective 1**) **Add and subtract rational expressions with unlike denominators**

Add.

▶ **Work with me.**

1. $\dfrac{3}{x} + \dfrac{5}{2x^2}$

⏸ To add or subtract rational expressions, we must have a _____ _____.

Add.

▶ **Work with me.**

2. $\dfrac{6}{x-3} + \dfrac{8}{3-x}$

Remember that $\dfrac{a}{-b} = \dfrac{-a}{b} = -\dfrac{a}{b}$.

Subtract.

▶ **Work with me.**

3. $\dfrac{y+2}{y+3} - 2$

Section 7.4 Adding and Subtracting Rational Expressions with Unlike Denominators

---

⏸ Adding or Subtracting Rational Expressions with Unlike Denominators

Step 1: Find the _____ of the rational expressions.

Step 2: Rewrite each rational expression as an _____ expression whose denominator is the LCD found in Step 1.

Step 3: Add or subtract _____ and write the sum of difference over the common denominator.

Step 4: _____ or write the rational expression in simplest form.

Subtract.

▶ Work with me.

4. $\dfrac{3a}{2a+6} - \dfrac{a-1}{a+3}$

Add.

⏸ Pause and work.

5. $\dfrac{x+8}{x^2-5x-6} + \dfrac{x+1}{x^2-4x-5}$

▶ Play and check.

Section 7.5 Solving Equations Containing Rational Expressions

Complete the outline as you view Video Lecture 7.5. Pause ⏸ the video as needed to fill in the blanks. Then press Play ▶ to continue. Also, circle your answer to each numbered exercise.

(**Objective 1**) **Solve equations containing rational expressions**

Multiplication Property

Multiply both sides by the same non-zero number.

Solve. Check your solution.

▶ **Work with me.**

1. $\dfrac{x-3}{5} + \dfrac{x-2}{2} = \dfrac{1}{2}$

⏸ **Pause and work.**

2. $\dfrac{2}{y} + \dfrac{1}{2} = \dfrac{5}{2y}$

▶ Play and check.

Be careful when there are variables in the denominator. Make sure the proposed solution does not make any denominator 0. If so, it is an <u>extraneous</u> solution.

Section 7.5 Solving Equations Containing Rational Expressions

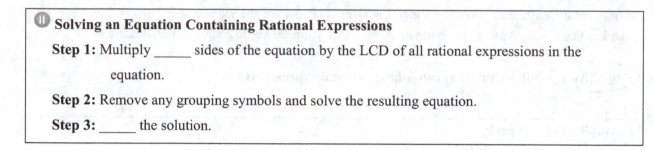

Solving an Equation Containing Rational Expressions

Step 1: Multiply _____ sides of the equation by the LCD of all rational expressions in the

equation.

Step 2: Remove any grouping symbols and solve the resulting equation.

Step 3: _____ the solution.

Solve. Check your solution(s).

Work with me.

3. $\dfrac{t}{t-4} = \dfrac{t+4}{6}$

Pause and work.

4. $2 + \dfrac{3}{a-3} = \dfrac{a}{a-3}$

Play and check.

Work with me.

5. $\dfrac{4r-4}{r^2+5r-14} + \dfrac{2}{r+7} = \dfrac{1}{r-2}$

Section 7.5 Solving Equations Containing Rational Expressions

$\left(\text{Objective 2}\right)$ **Solve equations containing rational expressions for a specified variable**

Solve for the indicated variable.

🔘 **Work with me.**

6. $T = \dfrac{2u}{B + E}$ for B

Section 7.5 Solving Equations Containing Rational Expressions

Section 7.6 Proportion and Problem Solving with Rational Equations

Complete the outline as you view Video Lecture 7.6. Pause ⏸ the video as needed to fill in the blanks. Then press Play ▶ to continue. Also, circle your answer to each numbered exercise.

Objective 1 **Solve proportions**

⏸ A _____ is the quotient of two quantities.

A _____ is a statement that two ratios are equal.

Cross Products

If $\dfrac{a}{b} = \dfrac{c}{d}$, then $ad = bc$.

Use cross products to solve for x.

▶ **Work with me.**

1. $\dfrac{x}{10} = \dfrac{5}{9}$

⏸ **Pause and work.**

2. $\dfrac{x+1}{2x+3} = \dfrac{2}{3}$

▶ Play and check.

⏸ A proposed solution that makes the denominator 0 is called an _____ solution.

Section 7.6 Proportion and Problem Solving with Rational Equations

---

(Objective 2) **Use proportions to solve problems**

Solve.

⏵ **Work with me.**

3. There are 110 calories per 177.4 grams of Frosted Flakes cereal. Find out how many calories are in 212.5 grams of this cereal. Round to the nearest whole calorie.

Find the unknown length x in the following pair of similar triangles.

⏸ **Pause and work.**

4.

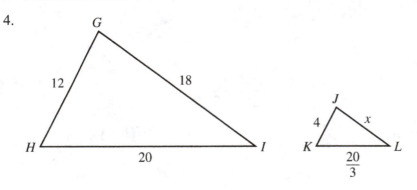

⏵ Play and check.

(Objective 3) **Solve problems about numbers**

⏵ **Work with me.**

5. Twelve divided by the sum of x and 2 equals the quotient of 4 and the difference of x and 2. Find x.

Section 7.6 Proportion and Problem Solving with Rational Equations

Objective 4 Solve problems about work

Work with me.

6. In 2 minutes, a conveyor belt moves 300 pounds of recyclable aluminum from the delivery truck to a storage area. A smaller belt moves the same quantity of cans the same distance in 6 minutes. If both belts are used, find how long it takes to move the cans to the storage area.

	minutes to complete job	part of job completed in 1 minute
belt		
smaller belt		
together		

Objective 5 Solve problems about distance

Work with me.

7. A car travels 280 miles in the same time that a motorcycle travels 240 miles. If the car's speed is 10 miles per hour more than the motorcycle's, find the speed of the car and the speed of the motorcycle.

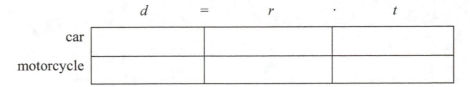

	d	=	r	·	t
car					
motorcycle					

Section 7.6 Proportion and Problem Solving with Rational Equations

Section 7.7 Variation and Problem Solving

Complete the outline as you view Video Lecture 7.7. Pause ⏸ the video as needed to fill in the blanks. Then press Play ▶ to continue. Also, circle your answer to each numbered exercise.

(**Objective 1**) **Solve problems involving direct variation**

Direct Variation Equations: a family of equations of the form $y = kx$

Direct Variation:

y varies directly as x, or y is directly proportional to x, if there is a non-zero constant k such that

$$y = kx$$

The number k is called the constant of variation or the constant of proportionality.

Linear Equation Form: $y = mx + b$; m is slope.

Write a direct variation equation that shows the relationship between the x- and y-values.

▶ **Work with me.**

1.

x	−2	2	4	5
y	−12	12	24	30

▶ **Work with me.**

2. y varies directly as x. If $y = 20$ when $x = 5$, find y when x is 10.

Section 7.7 Variation and Problem Solving

$\boxed{\text{Objective 2}}$ **Solve problems involving inverse variation**

⏸ _____ Variation: $y = \dfrac{k}{x}$

_____ Variation: $y = kx$

Inverse Variation

y varies inversely as x, or y is inversely proportional to x, if there is a non-zero constant k such that

$$y = \frac{k}{x}$$

The number k is called the constant of variation or the constant of proportionality.

Write an inverse variation equation that shows the relationship between the

x- and y-values.

⏵ **Work with me.**

3.

x	1	−7	3.5	−2
y	7	−1	2	−3.5

⏸ **Pause and work.**

4. y varies inversely as x. If $y = 5$ when $x = 60$, find y when $x = 100$.

⏵ **Play and check.**

With inverse variation, $y \cdot x = k$, the constant of variation.

$\boxed{\text{Objective 3}}$ **Other types of direct and inverse variation**

Direct and Inverse Variation as n^{th} Powers of x

y varies directly as a power of x if there is a non-zero constant k and a natural number n such that

$$y = kx^n$$

y varies inversely as a power of x if there is a non-zero constant k and a natural number n such that

$$y = \frac{k}{x^n}$$

Write an equation to describe the variation. Use k for the constant of variation.

▶ **Work with me.**

5. z varies directly as x^2

⏸ **Pause and work.**

6. y varies inversely as z^3

▶ Play and check.

▶ **Work with me.**

7. a varies inversely as b^3. If $a = \dfrac{3}{2}$ when $b = 2$, find a when b is 3.

Section 7.7 Variation and Problem Solving

$\boxed{\text{Objective 4}}$ **Variation and problem solving**

⬤ **Work with me.**

8. The distance a spring stretches varies directly with the weight attached to the spring. If a 60-pound weight stretches the spring 4 inches, find the distance that an 80-pound weight stretches the spring.

$d = k \cdot w$

Section 7.8 Simplifying Complex Fractions

Complete the outline as you view Video Lecture 7.8. Pause 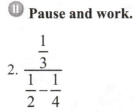 the video as needed to fill in the blanks. Then press Play to continue. Also, circle your answer to each numbered exercise.

Objective 1   **Simplify complex fractions using method 1**

A rational expression whose numerator or denominator contains fractions is called a _____

_____.

Simplify.

Work with me.

1. $\dfrac{-\dfrac{4x}{9}}{-\dfrac{2x}{3}}$

The <u>fraction bar</u> means division.

Pause and work.

2. $\dfrac{\dfrac{1}{3}}{\dfrac{1}{2} - \dfrac{1}{4}}$

Play and check.

Section 7.8 Simplifying Complex Fractions

> ⏸ **Method 1: Simplifying a Complex Fraction**
>
> **Step 1:** Add or subtract fractions in the numerator or denominator so that the numerator is a
>
> _____ _____ and the denominator is a _____ _____ .
>
> **Step 2:** Perform the indicated _____ by multiplying the numerator of the complex fraction
>
> by the _____ of the denominator of the complex fraction.
>
> **Step 3:** Write the rational expression in simplest form.

Simplify.

▶ **Work with me.**

3. $\dfrac{\dfrac{1}{5} - \dfrac{1}{x}}{\dfrac{7}{10} + \dfrac{1}{x^2}}$

(**Objective 2**) **Simplify complex fractions using method 2**

Simplify.

▶ **Work with me.**

4. $\dfrac{\dfrac{1}{5} - \dfrac{1}{x}}{\dfrac{7}{10} + \dfrac{1}{x^2}}$

> **Method 2: Simplifying a Complex Fraction**
>
> **Step 1:** Find the _____ of all the fractions in the complex fraction.
>
> **Step 2:** Multiply both the _____ and the _____ of the complex fraction by the LCD
> from Step 1.
>
> **Step 3:** Perform the indicated operations and write the result in simplest form.

Simplify the complex fraction using Method 2.

 Pause and work.

5. $\dfrac{\dfrac{x}{y}+1}{\dfrac{x}{y}-1}$

▶ Play and check.

Section 7.8 Simplifying Complex Fractions

Complete the outline as you view Video Lecture 8.1. Pause ⏸ the video as needed and fill in the blanks. Then press Play ▶ to continue. Also, circle your answer to each numbered exercise.

Objective 1 **Find square roots**

A number b is a square root of a number a, if $b^2 = a$.

$\sqrt{a}$ is the positive square root of a.

$-\sqrt{a}$ is the negative square root of a.

Find the square root.

▶ **Work with me.**

1. $\sqrt{49}$

▶ **Work with me.**

2. $-\sqrt{49}$

Square Root

If a is a positive number, then $\sqrt{a}$ is the positive square root of a and $-\sqrt{a}$ is the negative square root of a.

$\sqrt{a} = b$ only if $b^2 = a$ and $b > 0$.

Also, $\sqrt{0} = 0$.

Section 8.1 Introduction to Radicals

Find each square root.

⏸ **Pause and work.**

3. $\sqrt{36}$

▶ Play and check.

⏸ **Pause and work.**

4. $-\sqrt{16}$

▶ Play and check.

▶ **Work with me.**

5. $\sqrt{\dfrac{9}{100}}$

▶ **Work with me.**

6. $\sqrt{0}$

▶ **Work with me.**

7. $\sqrt{0.64}$

⏸ **Pause and work.**

8. $\sqrt{-4}$

▶ Play and check.

⏸ A square root of a negative number is _____ _____ _____ _____.

Objective 2 Find cube roots

Index $\longrightarrow$ $\sqrt[3]{8}$ $\longleftarrow$ radical sign

radicand

Find each cube root.

⏵ **Work with me.**

9. $\sqrt[3]{8}$

⏸ **Pause and work.**

10. $\sqrt[3]{125}$

⏵ Play and check.

⏵ **Work with me.**

11. $\sqrt[3]{-27}$

The cube root of a negative number <u>is</u> a real number.

Objective 3 Find n^{th} roots

Find each root.

⏵ **Work with me.**

12. $\sqrt[4]{81}$

Section 8.1 Introduction to Radicals

▶ **Work with me.**

13. $\sqrt[4]{-81}$

⏸ **Pause and work.**

14. $-\sqrt[5]{32}$

▶ Play and check.

▶ **Work with me.**

15. $\sqrt[5]{-32}$

⏸ Odd root of a negative number: _____ _____ _____

Even root of a negative number: _____ _____ _____ _____

(Objective 4) **Approximate square roots**

Approximate the square root to three decimal places.

▶ **Work with me.**

16. $\sqrt{136}$

(Objective 5) **Simplify radicals containing variables**

Simplify.

▶ **Work with me.**

17. $\sqrt{x^2}$; when $x = 3$

Section 8.1 Introduction to Radicals

For any real number a, $\sqrt{a^2} = |a|$.

Simplify. Assume that variables now represent positive numbers.

Work with me.

18. $\sqrt{x^4}$

Pause and work.

19. $\sqrt{36x^{12}}$

Play and check.

Work with me.

20. $\sqrt{\dfrac{x^6}{36}}$

Pause and work.

21. $\sqrt[3]{a^6 b^{18}}$

Play and check.

Section 8.1 Introduction to Radicals

Complete the outline as you view Video Lecture 8.2. Pause ⏸ the video as needed to fill in the blanks. Then press Play ▶ to continue. Also, circle your answer to each numbered exercise.

Objective 1 Use the product rule to simplify square roots

Product Rule for Square Roots

If $\sqrt{a}$ and $\sqrt{b}$ are real numbers, then $\sqrt{a \cdot b} = \sqrt{a} \cdot \sqrt{b}$.

Use the product rule to simplify.

▶ **Work with me.**

1. $\sqrt{20}$

▶ **Work with me.**

2. $\sqrt{33}$

⏸ **Pause and work.**

3. $\sqrt{180}$

▶ Play and check.

⏸ **Pause and work.**

4. $-5\sqrt{27}$

▶ Play and check.

8.2 Simplifying Radicals

Objective 2 Use the quotient rule to simplify square roots

Quotient Rule for Square Roots

If $\sqrt{a}$ and $\sqrt{b}$ are real numbers and $b \neq 0$ then $\sqrt{\dfrac{a}{b}} = \dfrac{\sqrt{a}}{\sqrt{b}}$.

Use the quotient rule to simplify.

⏵ **Work with me.**

5. $\sqrt{\dfrac{27}{121}}$

Objective 3 Simplify radicals containing variables

Simplify. Assume all variables represent positive numbers.

⏵ **Work with me.**

6. $\sqrt{x^{13}}$

⏸ **Pause and work.**

7. $\sqrt{81b^5}$

⏵ Play and check.

⏵ **Work with me.**

8. $\sqrt{\dfrac{12}{m^2}}$

$\boxed{\text{Objective 4}}$ **Simplify higher roots**

Product Rule for Radicals
If $\sqrt[n]{a}$ and $\sqrt[n]{b}$ are real numbers, then $\sqrt[n]{a \cdot b} = \sqrt[n]{a} \cdot \sqrt[n]{b}$.

Quotient Rule for Radicals
If $\sqrt[n]{a}$ and $\sqrt[n]{b}$ are real numbers and $b \neq 0$, then $\sqrt[n]{\dfrac{a}{b}} = \dfrac{\sqrt[n]{a}}{\sqrt[n]{b}}$.

Simplify.

⏵ **Work with me.**

9. $\sqrt[3]{250}$

⏸ **Pause and work.**

10. $\sqrt[3]{\dfrac{5}{64}}$

⏵ Play and check.

⏵ **Work with me.**

11. $\sqrt[4]{\dfrac{8}{81}}$

8.2 Simplifying Radicals

Section 8.3 Adding and Subtracting Radicals

Complete the outline as you view Video Lecture 8.3. Pause ⏸ the video as needed to fill in the blanks. Then press Play ▶ to continue. Also, circle your answer to each numbered exercise.

Objective 1 **Add or subtract like radicals**

⏸ **Like Radicals**

Like radicals are radical expressions that have the _____ _____ and the _____ _____.

Add or subtract.

▶ **Work with me.**

1. $3\sqrt{6} + 8\sqrt{6} - 2\sqrt{6} - 5$

⏸ **Pause and work.**

2. $4\sqrt{3} - 8\sqrt{3}$

⏸ **Play and check.**

▶ **Work with me.**

3. $\sqrt{11} + \sqrt{11} + 11$

$\sqrt{a}$ means $1 \cdot \sqrt{a}$ or $1\sqrt{a}$.

Section 8.3 Adding and Subtracting Radicals

Subtract.

⏵ **Work with me.**

4. $2\sqrt[3]{2} - 7\sqrt[3]{2} - 6$

(**Objective 2**) **Simplify radical expressions, and then add or subtract any like radicals**

Add or subtract by first simplifying each radical and then combining any like radical terms. Assume that all variables represent positive real numbers.

⏵ **Work with me.**

5. $\sqrt{12} + \sqrt{27}$

⏸ **Pause and work.**

6. $5\sqrt{2x} + \sqrt{98x}$

⏵ Play and check.

Work with me.

7. $\sqrt{\dfrac{3}{64}} + \sqrt{\dfrac{3}{16}}$

Pause and work.

8. $\sqrt[3]{8} + \sqrt[3]{54} - 5$

Play and check.

Section 8.3 Adding and Subtracting Radicals

Section 8.4 Multiplying and Dividing Radicals

Complete the outline as you view Video Lecture 8.4. Pause ⏸ the video as needed to fill in the blanks. Then press Play ▶ to continue. Also, circle your answer to each numbered exercise.

(Objective 1) **Multiply radicals**

Product Rule for Radicals

If $\sqrt[n]{a}$ and $\sqrt[n]{b}$ are real numbers, then $\sqrt[n]{a} \cdot \sqrt[n]{b} = \sqrt[n]{a \cdot b}$.

Multiply and simplify. Assume that all variables represent positive real numbers.

▶ **Work with me.**

1. $\sqrt{10} \cdot \sqrt{5}$

▶ **Work with me.**

2. $\left(6\sqrt{x} \right)^2$

⏸ **Pause and work.**

3. $\sqrt[3]{12} \cdot \sqrt[3]{4}$

▶ Play and check.

⏸ **Pause and work.**

4. $\sqrt{6}\left(\sqrt{5} + \sqrt{7} \right)$

▶ Play and check.

Section 8.4 Multiplying and Dividing Radicals

⏸ **Pause and work.**

5. $\left(\sqrt{x}+6\right)\left(\sqrt{x}-6\right)$

▶ Play and check.

Product Rule for Radicals

If a is a positive number, $\sqrt{a}\cdot\sqrt{a}=a$ or $\left(\sqrt{a}\right)^2=a$.

⟨ **Objective 2** ⟩ **Divide radicals**

Quotient Rule for Radicals

If $\sqrt[n]{a}$ and $\sqrt[n]{b}$ are real numbers and $b\neq 0$, then $\dfrac{\sqrt[n]{a}}{\sqrt[n]{b}}=\sqrt[n]{\dfrac{a}{b}}$.

Divide. Assume all variables represent positive real numbers.

▶ **Work with me.**

6. $\dfrac{\sqrt{90}}{\sqrt{5}}$

⏸ Pause and work.

7. $\dfrac{\sqrt{75y^5}}{\sqrt{3y}}$

▶ Play and check.

(**Objective 3**) **Rationalize the denominator**

⏸ _____ _____ _____ means rewriting a radical expression so that it does not contain a radical in the denominator.

Rationalize the denominator. Assume all variables represent positive real numbers.

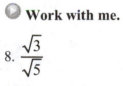 **Work with me.**

8. $\dfrac{\sqrt{3}}{\sqrt{5}}$

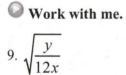 **Work with me.**

9. $\sqrt{\dfrac{y}{12x}}$

▶ **Work with me.**

10. $\sqrt[3]{\dfrac{1}{9}}$

Section 8.4 Multiplying and Dividing Radicals

Objective 4 **Rationalize using conjugates**

The conjugate of $a+b$ is $a-b$.
The conjugate of $a-b$ is $a+b$.

Rationalize the denominator.

Work with me.

11. $\dfrac{4}{2-\sqrt{5}}$

Section 8.5 Solving Equations Containing Radicals

Complete the outline as you view Video Lecture 8.5. Pause ⏸ the video as needed to fill in the blanks. Then press Play ▶ to continue. Also, circle your answer to each numbered exercise.

⬭Objective 1⬭ **Solve radical equations by using the squaring property of equality once**

⏸ Squaring both sides of an equation may cause _____ _____. Proposed solutions <u>must</u> be checked in the original equation.

The Squaring Property of Equality

If $a = b$, then $a^2 = b^2$.

Solve.

▶ **Work with me.**

1. $\sqrt{x+5} = 2$

▶ **Work with me.**

2. $3\sqrt{x} + 5 = 2$

⏸ **Pause and work.**

3. $\sqrt{x+6} + 1 = 3$

▶ Play and check.

207

Section 8.5 Solving Equations Containing Radicals

---

❶ Solving a Radical Equation Containing Square Roots

Step 1: Arrange terms so that one _____ is by itself on one side of the equation. That is,

_____ a radical.

Step 2: _____ both sides of the equation.

Step 3: _____ both sides of the equation.

Step 4: If the equation still contains a _____ term, repeat Steps 1 through 3.

Step 5: Solve the equation.

Step 6: Check all solutions in the _____ equation for extraneous solutions.

Solve.

⊳ Work with me.

4. $\sqrt{1-8x} - x = 4$

Objective 2 Solve radical equations by using the squaring property of equality twice

Solve.

⊳ Work with me.

5. $\sqrt{x-7} = \sqrt{x} - 1$

Section 8.6 Radical Equations and Problem Solving

Complete the outline as you view Video Lecture 8.6. Pause ⏸ the video as needed to fill in the blanks. Then press Play ▶ to continue. Also, circle your answer to each numbered exercise.

Objective 1 **Use the Pythagorean formula to solve problems**

> **The Pythagorean Theorem**
>
> If a and b are lengths of the legs of a right triangle and c is the length of the hypotenuse, then
>
> $$a^2 + b^2 = c^2$$

Find the length of the unknown side of the right triangle.

▶ **Work with me.**

1.

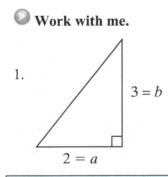

$3 = b$

$2 = a$

⏸ The _____ is the longest leg and lies opposite the right angle.

⏸ **Pause and work.**

2.

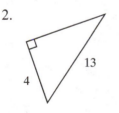

13

4

▶ Play and check.

Section 8.6 Radical Equations and Problem Solving

(Objective 2) **Use the distance formula**

Distance Formula

The distance d between two points with coordinates (x_1, y_1) and (x_2, y_2) is given by

$$d = \sqrt{(x_2 - x_1)^2 + (y_2 - y_1)^2}$$

▶ **Work with me.**

3. Find the distance between $(-3, 1)$ and $(5, -2)$.

⏸ **Pause and work.**

4. Find the distance between $\left(\dfrac{1}{2}, 2\right)$ and $(2, -1)$.

▶ Play and check.

Section 8.6 Radical Equations and Problem Solving

(**Objective 3**) **Solve problems using formulas containing radicals**

Work with me.

5. A wire is used to anchor a 20-foot-high pole. One end of the wire is attached to the top of the pole. The other end is fastened to a stake five feet away from the bottom of the pole. Find the length of the wire, to the nearest tenth of a foot.

Work with me.

6. The formula $v = \sqrt{2.5r}$ can be used to estimate the maximum safe velocity, v, in miles per hour, at which a car can travel if it is driven along a curved road with a radius of curvature, r, in feet. To the nearest whole number, find the maximum safe speed if a cloverleaf exit on an interstate has a radius of curvature of 300 feet.

.

Section 8.6 Radical Equations and Problem Solving

Complete the outline as you view Video Lecture 8.7. Pause ⏸ the video as needed to fill in the blanks. Then press Play ▶ to continue. Also, circle your answer to each numbered exercise.

Objective 1 **Evaluate exponential expressions of the form** $a^{1/n}$

Definition of $a^{\frac{1}{n}}$

If n is a positive integer and $\sqrt[n]{a}$ is a real number, then $a^{\frac{1}{n}} = \sqrt[n]{a}$.

Simplify.

▶ **Work with me.**

1. $25^{1/2}$

⏸ **Pause and work.**

2. $8^{1/3}$

▶ Play and check.

▶ **Work with me.**

3. $-16^{1/4}$

▶ **Work with me.**

4. $(-27)^{1/3}$

Section 8.7 Rational Exponents

⏸ **Pause and work.**

5. $\left(\dfrac{1}{9}\right)^{1/2}$

▶ Play and check.

$\boxed{\text{Objective 2}}$ **Evaluate exponential expressions of the form $a^{m/n}$**

Definition of $a^{m/n}$

If m and n are integers with $n > 0$ and if a is a positive number, then $a^{m/n} = \left(a^{\frac{1}{n}}\right)^m = \left(\sqrt[n]{a}\right)^m$.

Also, $a^{m/n} = \left(a^m\right)^{1/n} = \sqrt[n]{a^m}$.

Simplify.

▶ **Work with me.**

6. $32^{2/5}$

▶ **Work with me.**

7. $64^{3/2}$

⏸ **Pause and work.**

8. $8^{2/3}$

▶ Play and check.

Objective 3 Evaluate exponential expressions of the form $a^{-m/n}$

Definition of $a^{-m/n}$

If $a^{-m/n}$ is a non-zero real number, then $a^{-m/n} = \dfrac{1}{a^{m/n}}$.

Simplify.

▶ **Work with me.**

9. $625^{-3/4}$

▶ **Work with me.**

10. $-16^{-1/4}$

Objective 4 Use rules for exponents to simplify expressions containing fractional exponents

Simplify. Assume all variables represent positive real numbers.

▶ **Work with me.**

11. $3^{1/3} \cdot 3^{2/3}$

Section 8.7 Rational Exponents

▶ **Work with me.**

12. $\left(x^{2/3}\right)^9$

⏸ **Pause and work.**

13. $\dfrac{3^{-3/5}}{3^{2/5}}$

▶ Play and check.

Section 9.1 Solving Quadratic Equations by the Square Root Property

Complete the outline as you view Video Lecture 9.1. Pause ⏸ the video as needed to fill in the blanks. Then press Play ▶ to continue. Also, circle your answer to each numbered exercise.

⬭ **Objective 1** ⟩ **Use the square root property to solve quadratic equations**

⏸ A _____ _____ is an equation that can be written in the form

$ax^2 + bx + c = 0,\ a \neq 0$.

Zero Factor Theorem

If $a \cdot b = 0$, then $a = 0$ or $b = 0$.

Square Root Property

If $x^2 = a$ for $a \geq 0$, then $x = \sqrt{a}$ or $x = -\sqrt{a}$.

Solve each quadratic equation using the square root property.

▶ **Work with me.**

1. $x^2 = 64$

▶ **Work with me.**

2. $x^2 = -4$

⏸ **Pause and work.**

3. $2x^2 - 10 = 0$

▶ Play and check.

Section 9.1 Solving Quadratic Equations by the Square Root Property

▶ **Work with me.**

4. $(p+2)^2 = 10$

⏸ **Pause and work.**

5. $(3x-7)^2 = 32$

▶ Play and check.

(**Objective 2**) **Solve problems modeled by quadratic equations**

Solve the application using the formula $h = 16t^2$.

▶ **Work with me.**

6. The Hualapai Indian Tribe allowed the Grand Canyon Skywalk to be built over the rim of the
   Grand Canyon on their tribal land. The Skywalk extends 70 feet beyond the canyon's edge and is
   4000 feet above the canyon floor. Determine the time, to the nearest tenth of a second, it would
   take an object, dropped off the Skywalk, to land at the bottom of the Grand Canyon.

Section 9.2 Solving Quadratic Equations by Completing the Square

Complete the outline as you view Video Lecture 9.2. Pause ⏸ the video as needed to fill in the blanks. Then press Play ▶ to continue. Also, circle your answer to each numbered exercise.

Objective 1 Solve quadratic equations of the form $x^2 + bx + c = 0$ by completing the square

---

Square Root Theorem

If $x^2 = a$ (and a is positive), then $x = \pm\sqrt{a}$.

⏸ **Perfect Square Trinomial:** $a^2 + 2ab + b^2$ factors to _____.

Completing the Square

The number that completes the square on $x^2 + bx$ is $\left(\dfrac{b}{2}\right)^2$. To find $\left(\dfrac{b}{2}\right)^2$, find half the coefficient of

x, then square the result.

Solve by completing the square.

▶ **Work with me.**

1. $x^2 + 8x = -12$

⏸ **Pause and work.**

2. $x^2 - 2x - 1 = 0$

▶ Play and check.

Section 9.2 Solving Quadratic Equations by Completing the Square

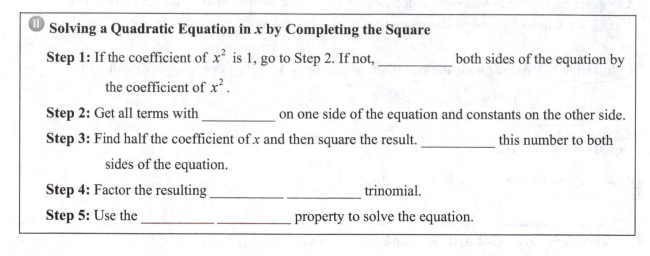

Solving a Quadratic Equation in x by Completing the Square

Step 1: If the coefficient of x^2 is 1, go to Step 2. If not, _____ both sides of the equation by the coefficient of x^2.

Step 2: Get all terms with _____ on one side of the equation and constants on the other side.

Step 3: Find half the coefficient of x and then square the result. _____ this number to both sides of the equation.

Step 4: Factor the resulting _____ _____ trinomial.

Step 5: Use the _____ _____ property to solve the equation.

Objective 2 Solve quadratic equations of the form $ax^2 + bx + c = 0$ by completing the square

Solve by completing the square.

Work with me.

3. $2y^2 + 8y + 5 = 0$

Section 9.3 Solving Quadratic Equations by the Quadratic Formula

Complete the outline as you view Video Lecture 9.3. Pause ⏸ the video as needed to fill in the blanks. Then press Play ▶ to continue. Also, circle your answer to each numbered exercise.

Objective 1 **Use the quadratic formula to solve quadratic equations**

A quadratic equation in standard form is $ax^2 + bx + c = 0$, where $a \neq 0$.

Quadratic Formula

If a, b and c are real numbers and $a \neq 0$, a quadratic equation written in the form $ax^2 + bx + c = 0$ has solutions

$$x = \frac{-b \pm \sqrt{b^2 - 4ac}}{2a}$$

Use the quadratic formula to solve each equation.

▶ **Work with me.**

1. $3k^2 + 7k + 1 = 0$

⏸ **Pause and work.**

2. $3 - x^2 = 4x$

▶ Play and check.

Section 9.3 Solving Quadratic Equations by the Quadratic Formula

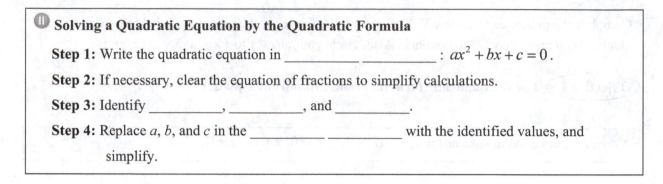

⑪ **Solving a Quadratic Equation by the Quadratic Formula**

Step 1: Write the quadratic equation in _____ _____: $ax^2 + bx + c = 0$.

Step 2: If necessary, clear the equation of fractions to simplify calculations.

Step 3: Identify _____, _____, and _____.

Step 4: Replace a, b, and c in the _____ _____ with the identified values, and simplify.

Use the quadratic formula to solve the equation.

▶ **Work with me.**

3. $5z^2 - 2z = \dfrac{1}{5}$

⬭ **Objective 2** Approximate solutions to quadratic equations

Use the quadratic formula to solve the equation. Find the exact solutions; then approximate these solutions to the nearest tenth.

▶ **Work with me.**

4. $x^2 = 9x + 4$

Section 9.3 Solving Quadratic Equations by the Quadratic Formula

⊙ **Objective 3** Determine the number of solutions of a quadratic equation by using the discriminant

Use the discriminant to determine the number of solutions of the quadratic equation.

▶ **Work with me.**

5. $3x^2 + x + 5 = 0$

⏸ $b^2 - 4ac$ is called the _____.

The following table corresponds the discriminant $b^2 - 4ac$ of a quadratic equation of the form $ax^2 + bx + c = 0$ with the number of solutions of the equation.

Discriminant

$b^2 - 4ac$	**Number of Solutions**
Positive	Two distinct real solutions
Zero	One real solution
Negative	No real solution

Find the number of real solutions without solving.

▶ **Work with me.**

6. $9x^2 + 2x = 0$

⏸ **Pause and work.**

7. $4x^2 + 4x = -1$

Section 9.3 Solving Quadratic Equations by the Quadratic Formula

Play and check.

Section 9.4 Complex Solutions of Quadratic Equations

Complete the outline as you view Video Lecture 9.4. Pause ⏸ the video as needed to fill in the blanks. Then press Play ▶ to continue. Also, circle your answer to each numbered exercise.

Objective 1 > **Write complex numbers using *i* notation**

The complex number system includes an imaginary unit, *i*.

Imaginary Unit *i*

The imaginary unit, written *i*, is the number whose square is -1. That is, $i^2 = -1$ and $i = \sqrt{-1}$.

Simplify.

▶ **Work with me.**

1. $\sqrt{-9}$

⏸ **Pause and work.**

2. $\sqrt{-63}$

▶ Play and check.

Complex Numbers and Pure Imaginary Numbers

A <u>complex</u> <u>number</u> is a number that can be written in the form $a + bi$, where a and b are real numbers. A complex number that can be written in the form $0 + bi$, where $b \neq 0$, is also called a <u>pure</u> <u>imaginary</u> <u>number</u>.

Section 9.4 Complex Solutions of Quadratic Equations

Objective 2 **Add or subtract complex numbers**

Add.

Work with me.

3. $(2-i)+(-5+10i)$

Pause and work.

4. Subtract $(2+3i)$ from $(-5+i)$.

Play and check.

Objective 3 **Multiply complex numbers**

Multiply.

Work with me.

5. $-9i(5i-7)$

Work with me.

6. $(4-3i)(4+3i)$

The conjugate of $a+bi$ is _____.

The conjugate of $a-bi$ is _____.

Objective 4 **Divide complex numbers**

Divide. Write the answer in the form $a + bi$.

▶ **Work with me.**

7. $\dfrac{7 - i}{4 - 3i}$

Objective 5 **Solve quadratic equations that have complex solutions**

Solve.

▶ **Work with me.**

8. $(x + 1)^2 = -9$

⏸ **Pause and work.**

9. $2m^2 - 4m + 5 = 0$

▶ Play and check.

Section 9.4 Complex Solutions of Quadratic Equations

Section 9.5 Graphing Quadratic Equations

Complete the outline as you view Video Lecture 9.5. Pause 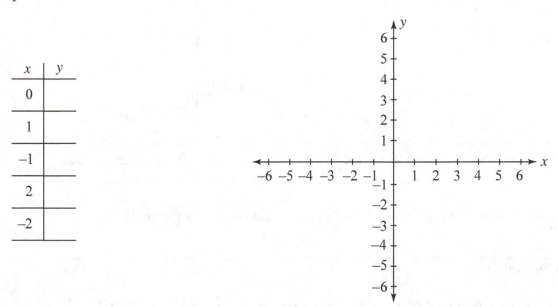 the video as needed to fill in the blanks. Then press Play to continue. Also, circle your answer to each numbered exercise.

Objective 1 **Graph quadratic equations of the form** $y = ax^2$

Quadratic Equation in Two Variables

$y = ax^2 + bx + c$; a, b and c are real numbers, $a \neq 0$

The graph of a quadratic equation in two variables is a _____.

Graph the quadratic equation by finding and plotting ordered pair solutions.

Work with me.

1. $y = 2x^2$

x	y
0	
1	
−1	
2	
−2	

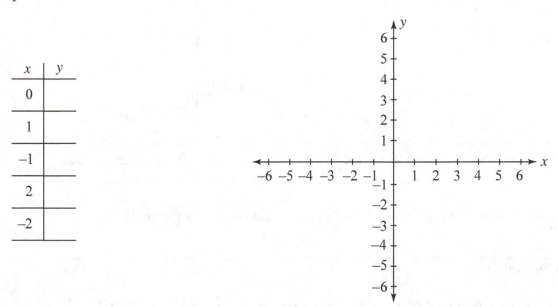

Parabola: opens upward, lowest point is vertex; opens downward, highest point is vertex.

Section 9.5 Graphing Quadratic Equations

Objective 2 Graph quadratic equations of the form $y = ax^2 + bx + c$

Graph the quadratic equation. Label the intercepts.

● **Work with me.**

2. $y = -x^2 + 4x - 3$

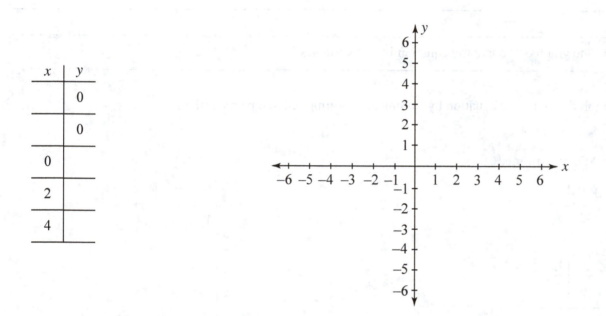

x	y
	0
	0
0	
2	
4	

⑪ For $y = ax^2 + bx + c$, if $a < 0$, the parabola opens _____. If $a > 0$, the parabola opens _____.

Objective 3 Use the vertex formula to determine the vertex of a parabola

Vertex Formula

The vertex of the parabola $y = ax^2 + bx + c$ has x-coordinate $\dfrac{-b}{2a}$.

The corresponding y-coordinate of the vertex is found by substituting the x-coordinate into the equation and evaluating y.

Section 9.5 Graphing Quadratic Equations

Graph the quadratic equation. Label the vertex and the intercepts.

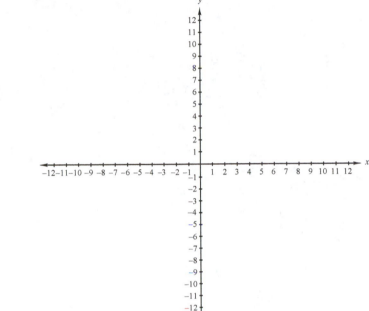 **Work with me.**

3. $y = 2x^2 - 11x + 5$

x	y

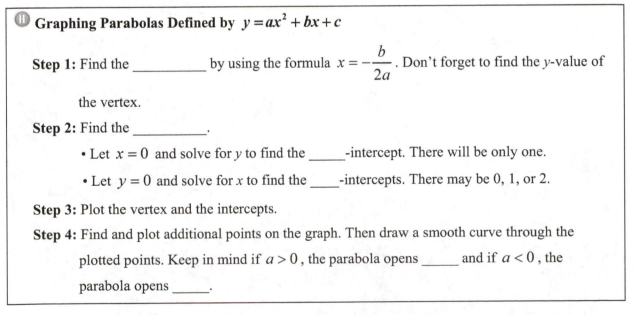

Graphing Parabolas Defined by $y = ax^2 + bx + c$

Step 1: Find the _____ by using the formula $x = -\dfrac{b}{2a}$. Don't forget to find the y-value of

the vertex.

Step 2: Find the _____ .

• Let $x = 0$ and solve for y to find the _____-intercept. There will be only one.

• Let $y = 0$ and solve for x to find the _____-intercepts. There may be 0, 1, or 2.

Step 3: Plot the vertex and the intercepts.

Step 4: Find and plot additional points on the graph. Then draw a smooth curve through the

plotted points. Keep in mind if $a > 0$, the parabola opens _____ and if $a < 0$, the

parabola opens _____ .

Section 9.5 Graphing Quadratic Equations